THE
CHANGE FUNCTION

THE
CHANGE FUNCTION

WHY SOME TECHNOLOGIES TAKE OFF
AND OTHERS CRASH AND BURN

Pip Coburn

PORTFOLIO

PORTFOLIO
Published by the Penguin Group
Penguin Group (USA) Inc., 375 Hudson Street, New York, New York 10014, U.S.A.
Penguin Group (Canada), 90 Eglinton Avenue East, Suite 700, Toronto, Ontario, Canada M4P 2Y3
(a division of Pearson Penguin Canada Inc.)
Penguin Books Ltd, 80 Strand, London WC2R 0RL, England
Penguin Ireland, 25 St. Stephen's Green, Dublin 2, Ireland (a division of Penguin Books Ltd)
Penguin Books Australia Ltd, 250 Camberwell Road, Camberwell, Victoria 3124, Australia
(a division of Pearson Australia Group Pty Ltd)
Penguin Books India Pvt Ltd, 11 Community Centre, Panchsheel Park, New Delhi – 110 017, India
Penguin Group (NZ), Cnr Airborne and Rosedale Roads, Albany, Auckland 1310, New Zealand
(a division of Pearson New Zealand Ltd)
Penguin Books (South Africa) (Pty) Ltd, 24 Sturdee Avenue, Rosebank, Johannesburg 2196,
South Africa

Penguin Books Ltd, Registered Offices: 80 Strand, London WC2R 0RL, England

First published in 2006 by Portfolio, a member of Penguin Group (USA) Inc.

10 9 8 7 6 5 4 3 2 1

PUBLISHER'S NOTE: This publication is designed to provide accurate and authoritative information in regard to the subject matter covered. It is sold with the understanding that the publisher is not engaged in rendering legal, accounting or other professional services. If you require legal advice or other expert assistance, you should seek the services of a competent professional.

PHOTOGRAPH CREDITS: Page viii: photograph of Sean Debow by Jennifer Debow; photograph of David Harvey by Susan Cook. Page xi: photograph of Duff McDonald by Carl Duquette; photograph of Melanie Wyld by Adam Gagas.

LIBRARY OF CONGRESS CATALOGING IN PUBLICATION DATA
Coburn, Pip.
    The change function : why some technologies take off and others crash and burn / Pip Coburn.
        p. cm.
    ISBN 1-59184-132-1
    1. Technological innovations. 2. Business enterprises—Technological innovations. 3. High technology industries. 4. Technology industry. I. Title.
T173.8.C62 2006
338.4'76—dc22   2005056471

Printed in the United States of America

Dedicated to Kelly, Bailey, Tucker, and Eamon

. . . the core of my life . . .

THANK YOU WITH PICTURES

"There is a theory which states that if anyone discovers exactly what the universe is for and why it is here, it will instantly disappear and be replaced by something even more bizarre and inexplicable. There is another theory which states this has already happened."

—Douglas Adams, *The Restaurant at the End of the Universe*

My own theory is that any technology book that hopes to become a great technology book should start with the quote above from Douglas Adams!

I'm what you'd call a global technology strategist. I have been so for seven years. But that "I" is really a "we." Technology strategy—particularly *global* technology strategy—is really a team game. There's just way too much ground for one person to cover. In various jobs, at various times, I have enjoyed the help of tremendous teammates during the last several years—Dave Bujnowski, Faye Hou, Fazia Amin, Diana Astuto, Pamela Maine, Qi Wang, Betty Wu, David Harvey, Sean Debow, Carolyn Curry, and Arlene Phillip. You have all treated me like gold.

I plan to continue what I'm doing for a long time to come, depending, I suppose, on whether anyone remains willing to listen to my hallucinations. I am fortunate each day to direct those hallucinations toward

Dave Bujnowski   Faye Hou   Fazia Amin

Diana Astuto   Qi Wang   Betty Wu and Adam Gallistel

David Harvey   Sean Debow   Carolyn Curry

Pamela Maine   Arlene Phillip

clients who use me to understand the world a bit better and to help them make better decisions about how to invest.

During the bulk of the nineties, I was a technology analyst and portfolio manager at Lynch and Mayer. It was a superb job. I was very lucky to receive mentoring from the likes of Ed Petner, John Levinson, and Eldon Mayer—all of who brainwashed me about looking for change. I thank you all.

In 1999, the folks at UBS were crazy enough to place me in the center of their global technology research practice—one hundred-plus analysts in twenty-plus countries. My job was to sit at the center of it all, to direct and synthesize. When I asked my soon-to-be bosses Stephen Carr and Raul Esquivel why they wanted to hire me, Stephen said, "Well, we think you'll get along with everyone around the world and you'll figure out the job along the way."

For six years, they gave me everything I could possibly hope for toward figuring it out—figuring out what would *change* next. UBS literally opened up the world to me and it was there where huge chunks of my thinking—whether you think it's any good or not—developed. I was born in Cleveland, Ohio, meaning I grew to be a humble Indians fan, always wanting the world but never expecting much. UBS helped a kid from Ohio become global and mainstream in the way Thomas Friedman might have described in writing *The World Is Flat.*

I could not be more thankful to UBS, not least for their support in creating this book. Thanks to Raul Esquivel, Erika Karp, and Gregg Goldman for helping make it possible. I have been fortunate that UBS gave me the opportunity to work daily with sharp analysts like Nikos Theodosopolous, Jeffrey Schlesinger, Russ Mould, Joe Dutton, Chitung Lui, Sharon Su, Byron Walker, John Hodulik, Aryeh Bourkoff, Mike Wallace, Adam Frisch, Heather Bellini, Gary Gordon, Ian McLennan, Mark Precious, and many, many, many others.

What I *really* do for a living is study change. I then apply what I know to my areas of domain expertise in technology, telecommunications, and media—all areas where change has been abundant. But what I really do is study change. It is a passion.

So what's in it for me is that I get to study change. And in one form or another I suspect I will be doing so for the rest of my life and hoping to use whatever I might learn to make the world a slightly better place than it would have been otherwise.

If Albert Einstein couldn't figure it all out, neither will I—which is perfect, since studying change becomes a never-ending pursuit. This book is a work in process. I plan to live until about one hundred and thirty, so at age forty, it is comforting to suspect that for the next ninety years I will never be bored. I won't accomplish my life's work anytime soon or, in fact, ever!

That's what is in it for me. But it gets better. In order to study change, I have had to continually be open to learning from everyone. I've been forced to expand the way I think about everything—to see things from entirely new perspectives than the one I currently had—and I must consciously seek out people who have pieces to this infinite puzzle and who expand my ability to think.

In common English, one might say I get to meet an awful lot of amazing people. The folks I'd like to thank have either inadvertently or explicitly helped me expand my thinking. Randy Komisar, Brian Arthur, Terry Pearce, Jerry Michalski, Melanie Wyld, Keith Yamashita, John Borthwick, Paul Pangaro, C. J. Maupin, Melinda Davis, John Dillon, Jeff Hawkins, DeWayne Hendricks, Esther Dyson, Douglas Engelbart, Et Perold, Steve Crandall, Arnie Berman, John Perry Barlow, Grant Mc-Cracken, John Maeda, Shelwyn Weston, Jaron Lanier, Joseph Tainter, Polly Labarre, Johna Johnson, Carl Johnson, Jennifer Corriero, Peter Cochrane, Len Kleinrock, Timothy Prickett Morgan, Dave Burstein, Paul Gustafson, J. C. Herz, Mike Hawley, Eric Hopkins, Kian Ghazi, Tim Harrington, Adam DeVito, Mary Hodder, Howard Greenstein, Joy Tang, David Isenberg, Andrew Odlyzko, Zachary Karabell, Eric Bonabeau, Tom DeMarco, Maureen O'Hara, Josh Wolfe, Rufus Winton, Kevin Ferguson, Om Malik, Andy Kessler, Kristin McGee, and certainly not last or least Ken Davidson who helps me see the world more clearly and has facilitated my best brainstorming.

I apologize that this list is ridiculously insufficient.

I am also thankful to Tony Perkins of *Red Herring* fame and now of AlwaysOn. Thanks to Nina Davis, my editor at AlwaysOn, and a ridicu-

Mel Wyld   Duff McDonald

lously large thanks to Duff McDonald who edited me with patience for several years when I was contributor to Red Herring and edited this book from cover to cover so the ideas would be more accessible to you. He is amazing in his craft. If the writing looks crisp, please don't confuse it for my own! Thank you to Adrian Zackheim and Adrienne Schultz at Portfolio for helping me realize a long-held dream to get my ideas about change to a broader and more divergent audience than typically reads my work. Thank you to David Kuhn as well for getting my thinking out further into the world. I greatly appreciate the help from all.

There are many writers who have expanded my thinking, but none more than Thomas Kuhn and his simple book, *The Structure of Scientific Revolutions,* written more than forty years ago. Little did I know that my "change" hero Thomas Kuhn was the uncle of my agent David Kuhn. Kismet.

I think it makes sense to thank John Granholm for suggesting I read more and more science fiction to expand my imagination. So, thanks as well for the works of Orson Scott Card, Neal Stephenson, Ray Bradbury, Robert Heinlein, John Brunner, Isaac Asimov, Vernor Vinge, and Douglas Adams.

Thank you also to the many folks at Reactrix and Salesforce.com who participated in the project and brought more life to the concepts within. I never want my thinking confused as monotonous impractical dribble.

Most importantly, thank you, Kelly, Bailey, Tucker, Eamon, Milan, Kai, Ralph, and Clarrie. Thank you also to Sandee, Ted, Diane, and Drew—you add so much to my life. Thank you, Dad. Thank you, Mom.

I never wanted to write a book if it meant immediately apologizing to the most important people in my daily life for being inaccessible and curmudgeon-like for the prior 127 months. I want my work and home to grow together. My family is the greatest thing I have to share and the center of all I do. One picture I carry with me is of my son Eamon at age six months staring dead on into the camera with his gigantic blue eyes. The caption I have imagined since then is . . .

"Dad, if you are away from me for even a second it better be because you are doing something you find meaningful."

Thank you for all of your support at every turn in my life and I hope you feel the same level of support from me at every moment as well. To describe just exactly how happy I am to spend my life with Kelly is utterly impossible. I try and fail every day. You are the most amazing person I have ever met.

On a practical basis, thanks to Jean-Jacques, Starbucks, The Wayfarer, The Port Bakery and Café, Borders, The Thornwood Diner, The Cape Porpoise Kitchen, and every other place that has allowed me to sit and think and write for long periods without insisting I consume a representative amount of food and gain a representative amount of weight along the way.

Finally, occasionally in one's life there might be something introduced that is so great that life changes for the better forever and you might pay homage to it every day.

The iPod is one of those things. I want to thank many of the artists who unknowingly helped in my writing this book: Ben Folds, Jimmy Buffett, Robert Randolph, Coldplay, Green Day, Shelly Fairchild, The Cranberries, Patty Loveless, Garth Brooks, Randy Travis, U2, Barenaked Ladies, the Beatles, Elvis Costello, No Doubt, India Arie, Sheryl Crow, REM, The Cars, Pink Floyd, James Taylor, Jonathan Coulton, The Marshall Tucker Band, the Allman Brothers, Nirvana, The La's; and when the going got tough either Alanis Morrissette or the Rolling Stones. When times got even tougher I resorted to my "go to"—Tom Petty and the Heartbreakers 1999 album *Echo*, which I am listening to right now.

The Coburn Family (*from left to right*): Eamon, Dani, Kelly, Tucker, Kai, me, and Bailey

Steve Jobs deserves credit as well for facilitating the easy and simple involvement of all these artists in the project without consequence of digital rights management.

> "*The significant problems we face cannot be solved at the same level of thinking we were at when we created them.*"
>
> —Albert Einstein

CONTENTS

INTRODUCTION

Early in my career, an old boss of mine, Ed Petner, insisted that I squeeze the entire investment case for every stock I ever wanted to buy onto a single sheet of paper. It was one of the most challenging tasks I've ever tackled, but it was well worth it. Hopefully, Ed will appreciate that the entire presumption of this book is whittled to just two pages—and I am even eating into it by relaying his direction. My aim is to address two key issues that exist in the technology industry today.

ISSUE 1: HIGH-TECH FAILURE RATES STINK

The commercial failure rate of nominally great new
technologies is troublingly high.
That failure rate is consistent with the hatred and distrust
most normal human beings—which I like to call Earthlings—
tend to have of high technology.
That hatred and distrust is a bummer since our little planet
can use all the help technology might provide.

ISSUE 2: SUPPLIERS THINK THEY ARE IN CHARGE BUT IN REALITY USERS ARE IN CHARGE

> The technology industry operates according to an implicit
> supplier-oriented assumption.
> That assumption is that if one builds great new disruptive
> technologies and lets cost reduction kick in,
> markets will naturally appear. This is known as
> "build it and they will come."

This mentality is a major problem. Adopting a new technology requires changing the habits of users. The industry acts as if change is easy when it's actually quite difficult. Users will change their habits when the pain of their current situation is greater than their perceived pain of adopting a possible solution—this is the crux of *The Change Function*.

I believe that users are *always* in charge and that supply is a necessary but not sufficient condition for commercial success. Companies and products geared toward this holistic user orientation will succeed at far greater rates than those stuck in a supplier-oriented mind-set.

The goal of this book is to look at what has failed in the past, to understand how the industry came to be in the position it is in today. And, through the prism of *The Change Function,* to spotlight examples of what might and might not work in the future, and to examine a few corporate cultures that seem to get it. But this is not eight easy steps to success. Change is *not* easy.

> *"It is not necessary to change. Survival is not mandatory."*
>
> —W. Edwards Deming

That's it. So long.

THE
CHANGE FUNCTION

ONE

SILICON VALLEY, WE'VE GOT A PROBLEM

> *"Nothing is more difficult than to introduce a new order. Because the innovator has for enemies all those who have done well under the old conditions and lukewarm defenders in those who may do well under the new."*
>
> —Niccolo Machiavelli, *The Prince*

We've all heard it before: "Build it and they will come." Well, the last six years have proven that at least in the technology industry, that maxim is shockingly—and expensively—untrue. But there's an alternative approach, one that is user oriented and not so supplier-centric. That's what *The Change Function* is all about.

A couple of years ago it dawned on me in an ah-hah moment that nearly every time tech company execs got in front of me in an effort to persuade me to invest my clients' money they were focused on themselves, what they had created, and why buyers would be smart enough to figure out how smart their technology was as the price came down. It was incredibly ethnocentric. It was all about the smart technologists and the "magic" that the smart technologists had created—their propositions were devoid of a deep understanding of what really went on in their users' minds. The alternative approach is for technology companies to become riveted to the needs and wants of the users they seek. It seems obvious when you say it out loud, but here goes: users are in charge of what they spend their money on—and they always have been. The technologists may be the magicians but the *users* have the checkbooks.

Questions from the audience?

So the user of new tech products has always been in charge?
Yes.

So we've been hallucinating that the vendors of technology are the more important part of the equation?
Yes.

Didn't David Moschella's book Customer-Driven IT *say the same thing?*
Kind of.

He said the customer is now *in charge . . .*
Yes, he did.

. . . and you say the customer has always *been in charge?*
Yes.

. . . and vendors were lured into believing they were in charge?
Yep.

. . . and during the 1990s it was easy to get confused and become supplier-centric?
Oh, yeah.

Because of the success of massive technology trends such as PCs and cell phones that had hit the mainstream?
I prefer "adopted" by the mainstream, but yes.

And now after the tech spending slowdown, the burst of the Internet bubble, and the near-80 percent drop of the Nasdaq stock index a couple of years back, the tech ecosystem is in crisis?
It sure is.

Because the supplier-centric orientation of tech companies and tech products no longer appears to be working?
Right.

And former Intel CEO Andy Grove's old mantra that the surest way to success is to focus on creating disruptive technologies that produce order-of-magnitude, or "10x," changes that dramatically alter the status quo seems so yesterday—so 1990s?
It does, doesn't it?

And no one cares so much about the importance of Gordon Moore's Law that price reductions on tech products can maintain a steady and aggressive pace and make cool technologies more and more affordable to ever large numbers of Earthlings . . .
True.

And no one quite knows what to do but create vague goofy hype phrases . . .
Ugh.

And they broadly talk of the Digital Home, OnDemand, and Grids . . .
Ugh.

And most of the companies are still stuck creating products that fail . . .
Ugh.

And this crisis provides an opportunity to suggest a new way . . .
Yes.

And really thinking about the user is a new concept even though everyone in the industry will claim that they have been doing the user-centric thing for years . . .
Pretty much.

And the tech industry thinks adopting new technology or facilitating change is easy.
They do.

Just build it and they will come. So there's been an over-reliance on Andy Grove's Law of 10x disruptive change and Gordon Moore's Law of cost reduction to make those wonderful disruptive technologies more and more affordable to more and more Earthlings.
Yes, there has.

And you and our friend Machiavelli think change is very difficult . . .
The two of us, plus many, many others.

And what you propose to do is to compare a potential user's need or crisis . . .
Keep going.

. . . with the total perceived pain of adopting the possible solution prior to forking over any money . . . a weighing machine with current pain on one side and perceived pain of the solution on the other . . .
Precisely.

. . . to see if change will happen.
That's it, in a nutshell.

THE GNAWING

"Now that we have progress so rapid that it can be observed from year to year, no one calls it progress. People call it change, and rather than yearn for it, they brace themselves against its force."
                                    —Stewart Brand, *The Clock of the Long Now*

As the Global Technology Strategist at UBS from 1999 to 2005, I led a group that generated nearly three hundred reports, mostly under the title The Weekly Global Tech Journey, that were aimed at synthesizing the worlds of technology and technology investing as best possible. We still do that at Coburn Ventures under the new title Waypoints. What we *really* do is aim to learn as much about *change* as we possibly can.

In the 1960s, Eldon Mayer—a key brainwasher of mine three decades later—went to the famed value investor Benjamin Graham and asked the Yoda of investing if he objected to his (young Eldon's) approach of looking for *change* as opposed to Graham's eloquently written investment volumes on looking for *value* as the key element in identifying promising stocks.

You see, Graham started with the assumption that the inherent value of an entity could be determined at any fixed point in time and that one needed only to observe whether the price of the stock in the public market was either more or less than that value. Graham became very famous for this simple logic. Graham told Eldon that *change* investing still counted as value investing. Whereas Graham looked for a differential between value and stock price at any particular moment, Eldon looked for shifts in value that would come about as a result of major changes in the life of the company over time. If Eldon could identify and understand these changes quicker and with greater clarity than others, he would *know* the new *value* before anyone else and could profit handsomely.

**With some fundamental changes at a company, the inherent
value of that entity changes *now* even if it takes most of
Wall Street a few years to really get it.**

With Yoda's blessing, Eldon ventured off into the investment world
and delivered spectacular results, even during the horrendous 1970s.
It wasn't until the early 1990s that I crossed his path. I joined his investment firm, Lynch and Mayer. I had studied change previously, but
now I would get to study change for a living. It seemed like heaven.
Here's where I got lucky: I got to combine *change thinking* with
the hottest, fastest changing portion of the investment world—
technology—and do so as part of a group led by superstar technology-
investment guru John Levinson. Technology is a great place to look
for change—and the money you can make from it—it's got the least
predictable business models and earnings estimates, and the greatest
stock volatility.

It wasn't easy. Luckily, after a handful of soul-searching episodes of
losing anywhere from 10 to 30 percent of a stock's value in a day, my
ability to recover from the knife-in-the-stomach reaction that often accompanies technology investing improved dramatically. I could again
sleep at night, I quit drooling and foaming at the mouth in public, and a
long-lasting peace among myself, change, and technology investing
emerged.

But none of this brings us back to why in July 2004, I found myself
sitting in a porch rocker, clicking on my laptop, and setting out a treatise
on *The Change Function*. Here's what happened. In our Weekly Global
Tech Journeys, we started pursuing a question that gnawed at us for
years:

Why does tech change?

More specifically,

Why in the world do new technologies get adopted?

Our goal: to develop an investment philosophy focused on recognizing patterns of change—as well as patterns where nothing changed—and to find the common thread in it all.

"PROBLEM? WHAT PROBLEM?"

"Despite the best efforts of remarkably talented people, most attempts to create successful new products fail. Over 60 percent of all new-product development efforts are scuttled before they ever reach the market. Of the 40 percent that do see the light of day, 40 percent fail to become profitable and are withdrawn from the market. By the time you add it all up, three-quarters of the money spent in product development investment results in products that do not succeed commercially."

—Clayton Christensen, *The Innovator's Solution*

In the above quote, Clayton Christensen—a guru of change inside enterprises—is being kind or generous or both. It seems somewhat hopeful to suggest that 25 percent of all technology research and development investment actually proves successful—that *only* 75 percent is a disappointment! My own sense is that 90 to 95 percent of new technology products fail to gain anything resembling the originally hoped-for success.

While much of this failure occurs at innumerable start-ups, lots of it also happens at the major corporate engines of the technology world. And as we see one of the most innovative tech companies on the planet—Apple—spending well below industry-wide norms on R & D as a percent of sales, it seems that effective and efficient R & D may have given way to sloppiness and accommodation across the technology industry during the five decades of success that followed the development of the transistor at Bell Labs in the 1940s.

The technology world—indeed, the entire technology ecosystem—accommodates such atrocious failure rates unnecessarily.

The implicit thinking that creating Andy Grove-style 10x disruptive technologies is an end in itself is not wrong but *limited*. By tolerating and accommodating pathetic success rates as opposed to examining the horrendous failure rates, the technology industry has generated numerous disservices, of which the primary one is as follows:

Technology is widely hated by its users. The potential for technology to prove beneficial for *everyone*—from creator to investor to the vast bulk of the planet's six billion residents—is currently being undermined.

> *"Technology is a queer thing. It brings you great gifts with one hand and it stabs you in the back with the other."*
>
> —C. P. Snow, British writer and scientist,
> *The New York Times*, March 15, 1971

THE RESPONSE

> *"Technology is not kind. It does not wait. It does not say please. It slams into existing systems . . . and often destroys them. While creating a new system."*
>
> —Economist Joseph Schumpeter

This remark from the famed economist Joseph Schumpeter typifies the relationship many in industry have with technology, including its creators. Schumpeter's quote might be translated to read, *Engineers create technology and the world is forever changed—no questions asked!* In that context, a desire to examine the horrendous success rates might be seen as an unwanted intrusion into the creative world that is the invention of new technologies.

"Huh?"

In studying change, it's hard to find actual people who would volunteer to be the ones studied—but the technology ecosystem consists of actual people. The ecosystem itself does not want to be studied either—just as most of us would react less than enthusiastically if our boss informed us that someone would be tracking our every move for the next few months to help find better ways of doing our job.

And so the occasional and unappreciated examination of failure in technology is met—as it is in most places—with defensiveness. That leaves us with a selective amnesia that looks for its lessons only from the winners—like the PC—but not from the losers—like the Picturephone. Didn't anyone ever stop to consider that the machismo-laden boast of the venture capital community—that their model *works well* when 90 percent of VC bets fail—might, just perhaps, be a little too forgiving? That it really is a little odd to celebrate the fact that only 10 percent of their bets work out in the end?

Ignore failure, and you can conclude that the system *does* work. That focusing solely on creating extraordinary 10x disruptive Andy Grove–style change and allowing Moore's Law to work its magic in lowering price so the wonders that technologists create can be made available to more and more Earthlings—the so-called price elasticity argument—is *all* that one must worry about.

The success of the PC proves it, they say! But I think they're wrong.

LESSONS OF THE PERSONAL COMPUTER

> *Alan Kay's job interview at Xerox PARC:*
> *"What do you think your greatest achievement will be at PARC?"*
> *"It'll be the personal computer."*
> *"What's that?"*
>
> —Michael Hiltzik, *Dealers of Lightning*

In the early 1970s, Alan Kay of Xerox PARC—the Palo Alto Research Center—called his personal computer a time machine. What he meant was that thanks to Moore's Law, the price would come down over time and a market would grow as a result. He was right—it did. Kay, a computer genius of legendary proportions, had seen the future in Gordon

Moore's April 1965 article "Cramming More Components onto Integrated Circuits" (*Electronics Magazine*, April 19, 1965).

Here's the issue, though: Moore's Law was a necessary but hardly a sufficient condition for the growth of the PC market. Sure, some otherwise very interesting markets would not have developed if the price point didn't drop to a level at which the products or services became economically palatable to potential customers. But with all due respect to flea markets, no one buys much of anything—whether it's an ugly shirt or a personal computer—just because the price is *really* cheap.

Alan Kay's genius was not simply that he waited around for the effects of Moore's Law to kick in but that he waited for the effects of the graphical user interface (GUI)—which was also developed at Xerox PARC—to kick in! Moore's Law might have brought the price points down on personal computers but the GUI made it accessible to millions and is primarily responsible for the massive growth of the personal computer market in the late 1980s and 1990s.

> *The GUI was the sufficient condition that made a cool technology less scary to Earthlings and therefore more accessible.*

But is that the lesson the industry learned? That you can make cool things, but you *also* need to make them less scary?

Nah!

In 1964, Wesley Clark built the first personal computer—LINC—at Washington University in St. Louis. In 1973, Cookie Monster crawled across a display at Xerox PARC in Palo Alto—on the Alto, the first recognizable personal computer by today's perception of what a PC is. In 1980, IBM introduced the PC and in 1981 it sold 65,000 of them. In 1989, just fewer than 20 million units were sold globally. Ten years later that figure approached 170 million.

The most famous meeting in the past fifty years of technology? Headquarters in Webster, New York, orders Xerox PARC in Palo Alto,

California, to give Steve Jobs and his merry men from Apple a no-holds-barred demonstration of the graphical user interface without having to sign a nondisclosure agreement. Xerox quite literally gave away that key technology in an afternoon. Later, Bill Gates and Jobs would battle over who owned the GUI, with Gates suggesting that Jobs was jealous of Gates who made more hay—or, more precisely, money—planting the technology in the Windows environment.

> *"Hey, Steve, just because you broke into Xerox's store before I did and took the TV doesn't mean I can't go in later and steal the stereo."*
>
> —Bill Gates, *Mac Week*, March 14, 1989

The most important giveaway in the history of the PC wasn't a chip or an operating system. It was a graphical user interface—that which would make the technology available to the masses without having them strain to learn how to write code. Steve Jobs's meeting at Xerox PARC gave him the key to lowering the total perceived pain of adoption for users.

The Change Function aims to identify the root of crisis by getting in users' heads as to what they *really* want—as opposed to running insightless focus groups—and it looks for ways of reducing the total perceived pain of adopting a new way of doing things. In other words, we want to understand the crisis at the adopter level, or specifically how a new offering solves a problem such that the pain in moving to a new technology is lower than the pain of staying in the status quo.

The Change Function
f (perceived crisis vs. total perceived pain of adoption)

Sometimes technologists forget just how vast the chasm is between them and *real people*. Many real people resent technology. So it won't be easy for technologists to survive this crisis intact—this realization that it is real people and not technologists who determine the fate of technologies. "Build-it-and-they-will-come"–thinking runs deep.

Long after its years of real growth evaporated, Intel still operates on the principle of self-cannibalization with a faith that the next use for the miracles they create will most certainly appear somewhere. Intel's CEO Andy Grove was named *Time*'s Man of the Year in the mid-1990s and was quoted as saying:

"Tech Happens"

There was no reference to users deciding when and if Tech Happens or, *our job is to help users easily adopt new technology,* but rather just that *Tech Happens.*

How 'bout this famous one from Arthur C. Clarke, author of *2001: A Space Odyssey:* "Any sufficiently advanced technology is indistinguishable from magic." Technologists must have *loved* Clarke for this engineering-centric comment, the implication of which is that Technologists=Magicians. Who wouldn't want to be considered a magician?

Granted, much of technology *can* seem indistinguishable from magic to most. But something needs to be added to the mix to create a business. The magician business, while quite exciting, is generally far from lucrative, with exceptions like David Copperfield who in addition to the magic itself adds the additional magic of figuring out a business model.

But change is difficult. For technologists, the focus today remains on building miracles and letting Moore's Law do the grunt work to create commerciality. The investment community adores miracles, overlooks failed miracles with astonishing ease, and finds total perceived pain of adoption far less alluring than jaw-dropping, gee-whiz technologies.

Supplier-centric adoption model = f (Andy Grove Law of 10× disruptive technology × Gordon Moore's Law)

At best, gee-whiz technologies are years ahead of schedule. At worst, they are mental indulgences lacking a potential user crisis. Yet gee-whiz tech continually grabs the headlines. And so we can rest assured that early next year the Sunday *New York Times*—among many others—will run a feature on smartphones for the seventy-fourth year in a row, despite the fact that they have been too early with this call for seventy-three years straight. There *won't* be a piece in that same Sunday *New York Times* about business intelligence software, because the average reader can't get their arms around what the heck it really is even though corporate customers keep consuming it at a predictably interesting pace.

Here's one of my 14,292 opinions on change and technology: the limited understanding of why technology is adopted has created a terribly misguided use of investment dollars and has likely sabotaged many a product that, if rethought, might have experienced far greater success . . .

> . . . and those failures aid and abet the hatred and distrust the world has for technology . . .

> and it doesn't work well for the world to hate and distrust technology . . .

> and anyone looking around right now might get the sense that the planet can use all the help technology might be able to provide.

You can be your own assessor as to whether this thinking is on the mark. Decide for yourself if technology is working:

- Are you tired of trying to remember power cords for business trips?
- Are you tired of trying to remember to charge your cell phone at night?
- Are you annoyed by your remote control—correction, are you annoyed by the seven remotes in your home because you can't find the one you really want?
- Are you annoyed by complex alarm clocks in hotel rooms?

- Are you annoyed because your back hurts when traveling, perhaps because of all that extra junk you're hauling with you in order to be connected?
- Does it stink when you don't know how to align spacing in a Word document and spend your day guessing how to fix it?
- Are you tired of being stuck on a helpline forever, bracing to hear a flurry of TechnoLatin when someone finally does answer?
- Are you annoyed when spell check Capitalizes words You don't Want capitalized?
- Are you worried about your privacy?
- Do you have too many passwords?
- Do dropped cell phone calls annoy you just a tad?
- Did you ever have a flashing 12:00 on your VCR?

The above list could go on forever—literally. Quite clearly, something is wrong. For all the miracles technologists are performing, there is still a lot getting in the way of it all just working, of the miracles being easier to understand and adopt, and for commercial success to follow.

THE DUMB SMARTPHONE

> "*A mobile phone needs a manual in the way that a teacup doesn't.*"
>
> —Douglas Adams . . . and he didn't mean a smartphone

In the midsixties, Fred Brooks helped develop the software that would instruct IBM's S/360 mainframes—a company-making event for IBM. In *The Mythical Man Month*, Brooks discusses the second-system syndrome, the main point of which is the following: In the first system, creators are happy to get the "system" out the door *anyway, anyhow*. And in the second system? Well, all the ideas rejected in the first system— with the aim of just getting the product to market—are chucked into version number two. All the fresh ideas dreamed up by the sales force are included as well, creating a "feature-laden," all-encompassing system that surely must be better! But it doesn't quite work that way. The second system is nasty, unattractive, expensive, and commercially unwanted. The word *bloatware* may not have existed when Brooks labored on the S/360, but that lone word is extremely descriptive.

Bloatware

The mobile phone industry—as a follow-up to its incredible success that was nearly a half century in the making—created extremely expensive, feature-laden, computeresque gizmos called smartphones. These phones have a variety of attributes, so the definition is quite fuzzy, but any phone allowing a user to run an Excel spreadsheet and featuring a Windows or Symbian operating system certainly qualifies as one.

Smartphones have continually undersold expectations, much to the dismay of technologists, while "bare bones" phones that accommodate nonserious and nonsmart activities such as $2 music ring-tone downloads will likely explode to nearly 800 million units sold in 2005.

But smartphones have garnered tremendous media attention because they are supposedly *smart* and represent the future or at least the future that some folks would like to see. It's great fodder for the Sunday *New York Times*.

Supplier Orientation: More is better.
Much more is much better.

User Orientation: More is confusing.
Much more is much more confusing.

Since users are in charge of their wallets, and much more confusion means a much greater sense of perceived pain of adopting a new technology, the odds of rampant purchasing are low. Time and time again, users reject a Swiss Army knife in nearly every facet of life except in a Swiss Army knife. The concept of the Swiss-Army-knife convergence gadget is both *ill thought* and a waste of time. Do we want an extension of function? Sure. But all-in-one convergence? Nah! The world isn't going toward one all-powerful device no matter what the common wisdom has been for the past decade. We have more devices than ever!

Smartphones provide a wonderful namesake for a repository of bloatware. Name the product *smart*, thus internally legitimizing jamming in *everything*. There suddenly is no such thing as a bad idea. It certainly makes logical sense to help consolidate devices—blah blah blah—except that the market for the smartphone gets whittled down to the technologists themselves, who live in a "more-is-better" world as opposed to "more is confusing."

The technology ecosystem hasn't *really* ever put two and two together. In fact, as I have written for AlwaysOn (www.alwayson-network. com) about the extremely low odds of an all-in-one gadget simplifying the world, I am continually met with nasty comments about how I am so *very, very* wrong and how I don't "get" how much better the world would be. I would love to be wrong, but for the past ten years such technologist dreams have been negated. Technology has proliferated into many forms as opposed to a logical single form and technology will likely continue to proliferate into even more forms.

I say that even though I would be among the ultimate beneficiaries. In May 2004 in Taipei, my friend Sean Debow and I laid out our gadgetry in one location to take a picture of all the backbreaking stuff we carry—I mean the wonderful technology we carry. I had twenty-seven pieces of gear myself. You might think Sean and I were perfect candidates for a smartphone—given the planetary inconveniences of our jobs—but *we* didn't even carry one. I see nothing noble, mind you, in lugging twenty-seven pieces of gear around on my back. In fact, my backache and the frustration of carrying the world on my shoulders around the world was what prompted the picture to begin with.

My fledging yoga practice is a response to the devices.

Today, in February 2005, I sit at the Port Bakery and Café in Kennebunkport, Maine, with my laptop, BlackBerry, cell phone, and iPod as standard operating support. Yesterday, I bought an iGo to hopefully reduce the stuff I must lug.

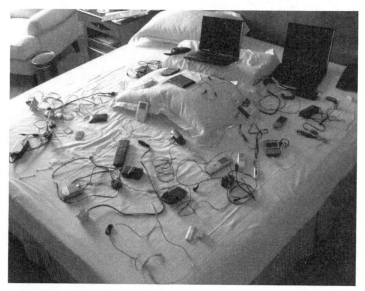

Pip's and Sean Debow's collection of necessary technology to stay connected, Taipei, 2004.

And now, one last flurry of frustrations like the finale of a fireworks show:

- Why do more expensive microwaves add features we never wanted and befuddle folks on basic usage? Do I really want to pay for a "popcorn" feature?
- Why do the lighting systems in four-star hotels require a potentially demeaning teach-in from the bellhop? Why should they be "systems" at all when light "switches" seemed to work pretty darn well?
- Why can't our friend Jerry at Goose Rocks Beach figure out how to operate his watch—lap 1, lap 2, mode 1, mode 2—to time his eight-year-old daughter running to the water and back?
- Why do we need to "operate" technology as opposed to just "use" it?
- Why do I have three remotes for my TV, DVD, and cable at my house in Maine, thus eliminating my desire to bother watching the muck on television anyway? Ahhh . . . that may be *smart*.

TWO

THE CHANGE FUNCTION

"And the day came when the risk to remain tight in the bud was more painful than the risk it took to blossom."

—Anaïs Nin

We've got a problem. Too often, products are designed, developed, and marketed by a technology culture that is supplier oriented to a fault. More often than not, products are created in a build-it-and-they-will-come mentality that relies solely upon Moore's Law for lowering prices and what can be called Grove's Law of generating 10x changes and improvements.

Grove's Law can be ascribed to anyone imploring others to build something that's remarkably better in some way, and Moore's Law can be ascribed to anyone suggesting that as price drops—voila!—a market will flourish. According to these arguments, *that's* why change happens in technology and why new products get adopted. Period. The thinking is that change in technology is a function of Grove's Law × Moore's Law.

Change = f (Moore's Law × Grove's Law)

I don't consider either Moore's Law or Grove's Law as wrong and I fully expect miracles to be created and prices to drop to make technologies available to those who can't afford the financial burden today.

Hopefully, those things will continue to happen faster and faster, but the thinking is limited if it starts and ends with the supplier.

> A single-minded focus on Grove's Law and Moore's Law creates a disservice only to the extent that it preempts or extinguishes more basic questions.
> **A Moore's Law fanatic might think of asking:**
> *"What will a user find important in this thing we are cost reducing and feature enhancing?"*
> **A Grove's Law fanatic might think of asking:**
> *"Have I created a tenfold improvement in actual user experience when I create a tenfold improvement in technological capability?"*

So here's the opportunity:

If the technology ecosystem realizes
that far greater success is just inches away
**that it's the user who decides whether or not to fork over the
money**
that the user has last say
always has
always will
**and if we want to obsess about something other than Moore
or Grove**
we might be well served obsessing about the potential users
and
not just the technology itself.

The supplier-centric technology world is in crisis. The solution is to switch the focus to a user-oriented model. The user-oriented Change Function says that change happens because of what is going on with the user—that the user has a weighing machine of perceived crisis versus the pain of adopting a different solution. If the crisis is less than the total perceived pain of adoption, there will be no change. If the crisis is greater than the total perceived pain of adoption, a change will occur.

> **So, Change** $= f$ (Moore's Law $\times$ Grove's Law)
>
> *is old and limited and supplier-centric, while*
>
> **The Change Function** =
> f (user crisis vs. total perceived pain of adoption)
> *is a new and fully explanatory user-centric framework.*

"So, let me see if I got this right . . . if the total perceived pain of adopting a smartphone is for whatever reason higher than the perceived crisis the user experiences without a smartphone, the user will not buy a smartphone. But if the crisis of not having a smartphone is greater than the total perceived pain of adopting the smartphone, then the user will buy a phone."

"Yep . . . you can see such a model in the everyday lives of individuals deciding whether to switch jobs, go to the gym, get married, go to rehab, or in their selection of a diet, and I don't think technology adoption is any different."

CRISIS EXPLAINED

"People hate change . . . and that's because people hate change. . . .
I want to be sure that you get my point. People really hate change.
They really, really do."

—Steve McMenamin, Atlantic Systems Guild

Back in 1962, Thomas Kuhn wrote what is still my favorite book on change: *The Structure of Scientific Revolutions.* Kuhn suggested that the entrenched interests in an old paradigm will resist change as long as they can until their own theories start to encounter massive holes in explaining the phenomenon at hand. Crisis precedes change. It takes crisis because change is difficult!

"There is no change without trauma."

—Bill Campbell, chairman of the board, Intuit

Even Einstein's theories, Kuhn noted, waited on hold until the gaps in Newtonian physics were being uncloaked. Imagine the resistance levels of physics professors around the world who might be out of date and out of a job if their knowledge base fell to pieces and upstart theories gained traction.

"Because it demands large-scale paradigm destruction and major shifts in the problems and techniques of normal science, the emergence of new theories is generally preceded by a period of pronounced professional insecurity. As one might expect, that insecurity is generated by the persistent failure of the puzzles of normal science to come out as they should. Failure of existing rules is the prelude to a search for new ones."

—Thomas Kuhn, *The Structure of Scientific Revolutions*

Even before I read Kuhn, the word *crisis* continually rattled through my head when I contemplated change so, I immediately felt at home with his arguments. I found it easy to extrapolate them to my world of investing in technology companies.

And so here I am on September 11, 2005, sitting on Delta's hip new Song Airlines flight DL 2006 heading to San Francisco. I have Ben Folds on my iPod—some of my most reliable writing music for whatever reason—and I am thinking again of Kuhn as I contemplate crisis in terms of *The Change Function*.

A couple of years back, as I was developing this framework with the folks on my team like Dave Bujnowski, Faye Hou, and Qi Wang, we kept stumbling our way on and around the word *crisis*. "It's a serious word," we thought. *It might distract listeners when the word pops out of my mouth.* Sure enough, it did seem to create a communication problem with those who were subjected to my babbling.

I found myself explaining, "Well, maybe not a full-fledged crisis but a problem or a serious desire." The problem was that people seemed to think I was suggesting that nothing ever changed without a heavy-duty *crisis* somehow involved. Even I didn't believe it in the simplest sense. Did I have a crisis prior to buying my first iPod? A *crisis*? Not really. But I did indeed buy one.

At breakfast eighteen months ago in Berkeley, Jaron Lanier, technology guru and friend to start-ups that often have trouble translating their gee-whiz technologies into businesses, simply said the obvious—that by crisis, I was describing a spectrum along which there might be various levels of crisis. I could draw a scale. When I think of crisis on one end of a spectrum relative to new products being offered up in the technology ecosystem I might think of "indifference" on the other end.

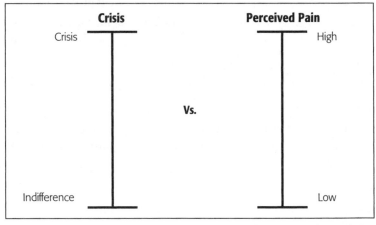

Source: UBS

That made a lot of sense. Potential users have some level of reaction from indifference to crisis whenever they encounter a new technology product. People are more apt to change—all else being equal—the higher the level of *crisis* that they have in their current situation.

What are some ways folks have *crisis* in technology? What do folks want so desperately they might consider their situation a crisis?

And here is where I run into trouble with some technology folks. Tech folks often still think users *should* have a problem and *should* decide that the technology offered is just way better than the status quo. Potential tech users *should* get that the technology being offered is much better than what they currently are using. That *should* be enough. Especially if the new solution has been deemed disruptive. But—and here's what I think is a very important point—even if a new tech solution is in fact disruptive by some technical assessment, it is often unclear to the users that such is the case regarding their own wants and needs.

And that can really get some folks in the tech ecosystem really mad! They sometimes revert to calling users *stupid*. A technology investor once said to me that, "People who can't use an iPod are stupid." I disagree. A few minutes later, I kept my mouth shut amid the irony when he noted that his mom would never get an iPod because she wouldn't be able to figure it out. I didn't tell his mom on him.

At a conference in September of 2005, a well-known—but to remain nameless—technologist calmly outlined in his Power Point presentation that technologists had to develop products for morons. Morons? No, nontechnologists are not morons unless a moron is someone unable to see the point of spending their free time fiddling with the innards of a personal computer. Stupid? No. Moron? Nah—just the opposite.

The view that nontechnical users are morons is pervasive in the technology ecosystem. I think this sentiment is part and parcel of the bitterness inside most corporate offices. The 95 percent of us who aren't IT guys think they *should* be able to get this stuff to work, be able to explain it in English, and know just a tad about the business of the company itself as opposed to just their own little corners of geekdom. And the reaction is that the 5 percent of personnel in IT label their accusers morons. The conflict may never end—but if you're in the business of wanting to sell technology to six billion folks on the planet, or even a mere billion, it would pay to give up the idea that potential users are wrong if they don't fawn over your offering.

My view? Sellers want to sell something. Potential users are "potential" until *they* decide to fork over money. There may very well be a *10x technological change* at hand but it may well be the case that it is *not* a 10x change in user experience and that is what matters.

The good news? Creating technologies does not have to be hit or miss. In the fall of 2005, Intel brought a real-live anthropologist on stage at its Intel Developer Forum to discuss how Intel understood the *user*. Finally, some progress. The whole thing suggests that Intel realizes that the users' desire to continually absorb greater and greater processor speeds has run out and those users may want some other things such as avoiding third-degree burns when using a laptop in their . . . well . . . lap. And so Intel has ended the gigahertz race in favor of other pursuits. Like, *What do users really want?*

When do users have a crisis? When all their friends have a flat-screen TV and they feel like dopes when admitting they don't. When their competitors' salespeople use both BlackBerrys and Saleforce.com's customer-relationship-management software and they don't—and it shows in the results, and their guys are getting hired away.

Who doesn't have a crisis? The guy already satisfied with his cable modem service who's being bombarded with advertisements to switch over to a me-too DSL service. The store that uses bar codes that work pretty well, even if they don't provide the exactness of precisely which tube of toothpaste they just sold, which is the promise of bleeding-edge, radio-frequency identification, or RFID.

I suspect that anthropologists will side with professor Ted Levitt, who long ago wrote in a 1960 *Harvard Business Review* piece titled "Marketing Myopia" that when people buy quarter-inch drill bits, it's not because they want the bits themselves. People don't want quarter-inch drill bits—they want quarter-inch holes. What he meant was that everything is a *service* attending to a need. When I think of the needs a certain *service* might provide I want to think it through pretty darn deeply.

An example: When someone buys and uses a cell phone the industry scores one up for "wireless." But no one buys "wireless." "Wire-less" tells us only what they didn't buy: they didn't buy something with wires. But what *service* might something without wires get you? The service is mobility and a type of freedom. When you buy *wireless*, you're buying freedom. Do people have a seemingly unquenchable desire for freedom? It would seem so.

A great technology *product* would be one that registers easily with the potential user as satisfying an underlying need such as freedom or communication or integration. There are many services that great technology might supply. There are many core wants that are far along on the continuum from indifference to crisis.

Sociologists observe people to see the connection between their habits and their wants. They don't ask. They aren't so big on surveys and they have a limited use for focus groups. They observe. They're on the lookout for crisis.

THE TOTAL PERCEIVED PAIN OF ADOPTION

Why do we open our wallets and decide to buy something, anything? What takes us from being *potential* customers to *actual* purchasers of a product or service?

Most people, technologists included, will tell you that's an easy question. *When the price is right,* they'll say. After all, we all—whether we will admit it or not—feel a little bit of pain when we take out that $20 bill (or write that $20,000 check)—and hand it over to someone else. Reduce that pain—make it $19.95 (or $19,995)—and we're more likely to fork it over, aren't we? In other words, a reduction in price reduces our total perceived pain of adoption.

Isn't that all there is to it?

Not at all.

Much of the time—and, more importantly, *most* of the time in technology purchases—price accounts for less than 10 percent of the *total* perceived pain of adoption, or TPPA.

I'll get to the other parts of the equation in a minute, but first a quick aside: it's an occasion to be flat-out thrilled—all of us, buyers and sellers included—if cost is the biggest portion of the total perceived pain of adoption. Thrilled.

Why? Because Moore's Law has been doing a good job throughout technology—to be clear, Gordon Moore's Law dealt exclusively with how price and performance of semiconductors changes over time—but throughout technology, prices come down as a rule. So, if you're *really* only reliant on cost reduction in order to make that sale, you're in pretty darn good shape as long as you have a bit of patience.

So, what are the other portions of TPPA? We've all endured them. The pain of them, that is. Reading instruction manuals, researching product information, feeling stupid when trying to learn or install a new gadget, experiencing the inability to find a salesperson that can explain the product in English, asking an eight-year-old kid for help who responds by saying, "Duh!" The list goes on.

The total perceived pain of adoption includes the long wait in the line to buy the product and the longer line waiting to talk to the help desk. It includes talking to the expert who assists you in feeling like an idiot for not knowing what RAM is. The pain involves learning something either slightly or radically new—a huge roadblock for people considering buying new technologies. Education is painful. Anything that might conceivably cause pain—no matter how small or how large—is a perceived pain of adoption. I use the word *total* to emphasize that the pain is much, much more than just the price itself.

But there's more. As any good marketer knows, "perception" is as much a part of any sale as reality—thus the inclusion of the word *perceived* in the TPPA. Before someone actually chooses to turn over money and adopt a technology, there is no *actual* pain of adoption because the adoption process hasn't started. All one has is the perception of how painful it will be to actually adopt the product and change a habit—how much trauma will be involved in the change. It is a *perception* about the future.

The TPPA is the sum of all these *a priori* considerations.

A friend of mine, Melanie Wyld, said she understood what the heck I was yakking about with all this total-perceived-pain-of-adoption stuff one day at an investor lunch in New York City. Bill Clinton was set to speak, so Mel got there early to get a spot. This was at the point Bill Clinton had just left office and set out on his way to pick up a couple hundred thousand dollars to talk for forty-five minutes—a great gig if you can get it. So Mel sat down at a table filled—not surprisingly—with men between the ages of forty and fifty-five. Unfortunately, it's still a male-dominated investment world.

Mel noted that the eight or nine men at her table—all awaiting Bill Clinton's arrival—weren't even talking the slightest bit about Clinton. Not a peep. They were otherwise engrossed. The topic was the Atkins Diet. Imagine—every man at that table was on the Atkins Diet! A bunch of middle-aged men sitting around talking about their diet. Mel thought that this was *amazing*. Had the world gone mad? She rattled off a four-page note to me about the incident. And it was all about TPPA!

She had determined that the success of the Atkins Diet—which led ultimately to the ridiculous situation of middle-aged men chatting about carbohydrates (or a lack thereof) while awaiting the arrival of the former President of the United States had to do with TPPA. These men were no more overweight than a similar group might have been five years previously. Their "crisis" hadn't increased at all. What the Atkins Diet did was reduce the pain of adoption. Middle-aged men no longer had to eat birdseed if they wanted to lose weight. They could eat meat and eggs. They could be real men—and lose weight—and not be embarrassed while talking about the number of pounds they had lost.

I still have no idea what Clinton spoke about that day.

Anyway, bringing it all back to the world of technology and why some technologies win while most lose—the Change Function looks at individual experiences on both sides of the equation. On the one side, it looks at the level of crisis with regard to the *services* a new technology product might offer. On the other, it looks at the total perceived pain of adoption associated with that new *service*. It's a weighing machine—not the type for a middle-aged man on the Atkins Diet to gauge his progress, but a weighing machine of a more abstract sort.

he level of crisis is higher than the total perceived pain of adopting / solution, then a change will occur. If the crisis is lower than the total perceived pain of adoption, then things will stay as they are.

> *"Every product according to its maker is easy to use. And for the folks who designed it, I'm sure it is. But, say what you will, so many products that should be easy to use, just aren't. This 'exasperation factor'—the sigh you heave when you hit that brick wall—undoubtedly causes many would-be buyers of a demo package just to leave, never to return. Still, vendors seem to be blissfully unaware of the opportunities that they are losing."*
>
> —Kevin Tolly, *Network World* magazine, July 4, 2004

THREE

IS THE TECHNOLOGY INDUSTRY READY FOR CHANGE?

"By now it's perfectly clear that the national fascination is riveted on people who are getting rich quickly and easily—and that it's hard for the country to see beyond the dollar signs. The Internet is the plot device for the 90s; it's the thing people are using to get rich, like oil and real estate in the 70s, or stocks and bonds in the 80s."

—Po Bronson, *The Nudist on the Late Shift*

Is there a crisis in the technology industry? Sitting here in late 2005, the question seems a little goofy. Yes, tech is in crisis. In fact, the technology ecosystem has itself come around toward acceptance of that fact. But it took awhile.

DENIAL | ANGER | BARGAINING | DEPRESSION | ACCEPTANCE

Source: UBS

If the question had been asked just a few years back, it would have prompted responses you can find in psychologist Elizabeth Kubler Ross's *On Death and Dying,* because the global technology ecosystem—particularly the part in the northern portion of California—was still in deep denial that anything had structurally changed.

Some folks got it. Folks like former COO at Oracle, Raymond Lane, who, at the 2004 Open Source Business Conference, said:

> *"This is a young industry, of young people. Half this industry is influenced by what happened in the 1990s. What we are experiencing now is normalcy, it's not a trough or recession. We can have melancholy, but designing a business around it is folly."*

Or retiring Microsoft CFO John Connors:

> *"We are in an era today when technology isn't really different from any other industry."*
>
> —As quoted in *The McKinsey Quarterly*

Ray Lane's former boss Larry Ellison was loudest of all and decided that an industry-wide consolidation was appropriate, a conclusion that prompted him to buy PeopleSoft:

> *"It's not coming back. There's this childlike notion that the IT industry is different than the auto industry or the railroad industry. That IT will always be immature, it will always be run by venture capitalists up on Sand Hill Road, by clever entrepreneurs with the Next Great Thing. And if you dare not agree with them, they say you just don't get it. Get what? The industry is maturing. The valley will never be what it was."*
>
> —As quoted in *Barron's,* January 25, 2003

I could go on with examples of people who saw the change and didn't resist it, but most of the troubled technology ecosystem was in the pain associated with resisting the change that was beating it down.

Here in the back half of 2005, the broad-based denial is gone and people have largely moved through at least a couple of Kubler Ross's stages and are much more open minded about the possibility that the industry is in crisis.

What follows is a quick glimpse of how high tech rose in order to see how much it fell, how quickly it fell, and the window that has been opened for change throughout the ecosystem. The window is now. This book is not meant as a lovely intellectual voyage but rather as a practical framework for actually identifying and implementing change.

> *"The significance of crisis is the indication they provide that an occasion for retooling has arrived."*
>
> —Thomas Kuhn, *The Structure of Scientific Revolutions*

> *"That era of Moore's Law where I get all three for free—transistors, power, and frequency—is over."*
>
> —Pat Gelsinger, Intel

> *"Speeds and feeds are just not that important anymore."*
>
> —David Nardi of the Yankee Candle Company,
> as quoted in *eWEEK*, June 28, 2004

Why do we even care if there is a crisis in technology? Well, if there is one, the industry is in great shape for change. Ideas that would have been shot down ten years ago will now gain an audience.

> *". . . these examples share another characteristic that may help to make the case for the role of crisis impressive: the solution to each*

of them had been at least partially anticipated during a period when there was no crisis in the corresponding science; and in the absence of crisis those anticipations had been ignored."

—Thomas Kuhn, *The Structure of Scientific Revolutions*

Not only have share prices and earnings collapsed and layoffs been stunningly large but all this came from a spectacularly high perch making the distinction between now and then extraordinarily obvious.

Technology has been a fabled story stemming from—geez, as far back as you wish—the discovery of fire? Certainly at least as far back as the 1947 invention of the transistor at Bell Labs or the 1958 development of the integrated circuit by Jack Kilby at Texas Instruments. The component semiconductor industry—a feeder into a wide range of end markets—experienced 17 percent compound annual sales growth between 1965 and 2000. We have experienced digital computing, the mainframe, minicomputers, and the PC revolution. We have experienced phenomena such as Maxwell's Rainbow, Marconi's cross-Atlantic transmissions, Edison's lighting and phonograph, Alexander Graham Bell's telephone, lasers and fiber optics, and the 1973 invention of the modern cell phone by Martin Cooper.

In 1961, the U.S. Department of Defense allocated $6 million for ARPA (the Advanced Research Projects Agency) to pursue a "command-and-control research project" that became ArpaNet, which became—ta-dah—the Internet. We've invented application software and gaming software, and we've experienced the unfortunate spawning of viruses. It's hard to know where to begin when discussing the changes information and communication technology have helped generate.

Technology spread from a geek-to-geek interaction when IBM delivered its Series 360 Mainframe that Fred Brooks created software for in 1965 to a geek-to-many conversation moving to a geek-to-every one of the six billion humans populating our little planet today. That's a big change. Heck, in 1965 could you have found a thirty-nine-year-old male UBS executive hiding out at the Port Bakery and Café in Kennebunkport telecommuting, let alone—and here's a tough one to fathom—"typing" on a "laptop"—let alone typing on anything if anyone remembering that artifact known as the typewriter is reading? Men didn't eat quiche and they didn't "type" . . . egads!

Times have changed *just* a wee bit.

And in the 1990s we saw monstrous changes.

Monstrous

The PC hit mainstream in the 1990s, nearly three decades after the first PC—the LINC—was created in 1964, and a full fifteen years after Bill Gates dropped out of Harvard. In 1989, less than 20 million PCs were sold globally. Ten years later the figure was closer to 170 million.

- The wireless revolution of the nineties led to today's count of nearly two billion global wireless subscribers.
- The database hit mainstream in the nineties.

- Enterprise Resource Planning software hit mainstream in the nineties.
- Customer Relationship Management software hit mainstream in the nineties.
- The Internet was deregulated in the nineties, allowing exposure to the masses. Executives cobbled together barely thought-out Internet strategies and parents bought PCs so their kids wouldn't be on the wrong side of the digital divide.
- The Y2K fix was fixed, whether it needed it or not.

And so on.

The Internet was such a big change from the status quo that even Bill Gates nearly missed it. In fact, if you go on eBay or Amazon and get a copy of Gates's book *The Road Ahead* in hardback, you will find little mention of it at all. Only in the paperback version did he deem a few more mentions of this Internet thing necessary to offer readers a better view of *The Road Ahead*.

But all the success was far from easy and was by no means overnight. It was decades in the making. Vannevar Bush wrote about many of these trends in his 1945 *Atlantic Monthly* piece "As We May Think." You read that right. Yes, 1945. "As We May Think" makes my list of the ten most significant technology pieces written during the twentieth century. With World War II over, Bush suggested that the United States might want to redirect engineering talent toward the next set of key technological issues. He dreamed up what amounted to the personal computer, storage, the Internet, and the World Wide Web. That was in 1945. The underlying point: tech didn't just "happen." We've only gotten where we are after a good long struggle.

The first mobile phone service in the United States was
available in 1946.
The first video game was played in 1961.
The first PC was built in 1964.
The first e-mail was sent in 1971.

Many of the amazing mass trends of the 1990s took decades to reach mainstream adoption, were unexpectedly facilitated by one-time regulatory changes, or both.

> *"It takes society thirty years, more or less, to absorb a new information technology into daily life. It took about that long to turn movable type into books in the fifteenth century. Television, invented in the mid 1920s, took until the mid-1950s to bind us to our sofas."*
>
> —Robert Cringely, *Accidental Empires*

The practical result of all this good news: From 1993 through 2000, technology companies grew earnings by a compound average growth rate of 25 percent—nearly twice any other sector—and in seven of those eight years, technology was ranked either first or second for earnings growth of all ten sectors. An awful lot was going on in the nineties. More importantly, the nineties *felt* darn easy in the technology ecosystem.

That was all good news.

If Thomas Kuhn were looking for a *crisis* to rethink the base culture of technology or to shed light on the 90 percent commercial failure rate of new products, the 1990s was precisely the wrong decade in which to be looking.

I wasn't spending too much time at *that* time thinking about the elephant head on the table—*the ridiculous failure rates that VCs were sponsoring while hunting for giant elephants*—because it really was a lot of fun participating in Nasdaq's run from 333 on October 10, 1990, to the 5,008 peak on March 10, 2000. That's 31 percent annual growth in share prices. No one wanted it to end. There is a famous bumper sticker you may have seen that reads:

"Please, God, just one more bubble."

PRODUCT	INVENTOR	YEAR
The 1st Video Game	MIT student Steve Russell creates Spacewar, the first interactive computer game, on a Digital PDP-1 mainframe computer.	1961
The 1st Internet	ARAPANET is formed to connect four major computers in southwestern U.S.	1969
The 1st Mouse	Douglas Engelbart invents an "x-y position indicator for a display system."	1970
The 1st E-mail	Computer engineer Ray Tomlinson sends a simple text message to himself.	1971
The 1st Ethernet	Bob Metcalfe creates Ethernet to connect computers at Xerox PARC. He started 3Com six years later.	1973
The 1st Personal Computer	The Altair is the first "personal computer" produced in relatively high quantity.	1975
The 1st TCP/IP	Early pioneers of computer networks agree upon a standard for TCP/IP.	1977
The 1st Cell Phone	The first commercial cellular phone system begins operation in Tokyo.	1979
The 1st World Wide Web	In an effort to make the Internet more user-friendly, CERN's European Organisation for Nuclear Research invents the World Wide Web.	1991
The 1st Linux System	Twenty-one year-old Linus Torvalds is inspired by the free software movement and creates Linux.	1991

Source: UBS

The positive disposition toward technology turned maniacal. Those who didn't pretend to "get it" were afraid of being accused of . . . well . . . not "getting it." The peer-to-peer pressure to embrace technologies grew extreme. As many PDAs were purchased to avoid cocktail party ostracism as were purchased to . . . well . . . what were they used for again? Oh, yeah—$500 address books. The general media finally heard about Moore's Law by 1999, successfully misquoted it, and conceptualized it as the rational explanation for the ever-expanding technology ecosystem. The whole thing would clearly go on forever. All this occurred *just* as it was all coming to an end. Technology gurus who would tell us what our lives would look like whether we wanted to know or not got in for fifteen minutes of fame just under the wire.

The fall guys just so happen to have been among the most optimistic. On Wall Street, we find them in Internet guru Henry Blodget and poster child for telecommunications excess Jack Grubman. George Gilder holds the same place in the "futurist" community. Gilder was just a normal man—an educator at heart until he got the technology bug and decided that technology would prove the greatest force in education. So he devoured all he could in a preparation to write his classic study of the semiconductor industry, *Microcosm*. The problem is that no one will remember George Gilder as the historian of the semiconductor age but rather as the guy who bought the vision of the future handed down by the most shamelessly unabashed hucksters of the *Telecosm* just as we were hitting the peak. If Tony Perkins wrote the book on why the Internet Bubble would burst, George Gilder is perhaps unfairly remembered as having written the book, *Telecosm*, explaining why the market—and his stock picks—were set to go much higher. Needless to say, he was wrong.

To be clear, George Gilder was not a flavor of the day. George Gilder represented the flavor of the decade. A discussion of tech in the nineties is simply incomplete without George Gilder.

> **The tech ecosystem was spoiled silly for fifty-plus years.**
> **The spoilage was particularly intense in the 1990s.**
> **Understandably, incredibly bad habits developed.**
> **Grove's Law x Moore's Law dominated!**
> **Nothing widely criminal . . . just human nature at work!**

> *"Betting on Internet stocks, it's actually pretty easy. Here are the rules: 50 bucks a share is cheap; $150 a share is fairly valued; but at $200 a share, the stock is cheap again because that means you are about to have a four-for-one split."*
>
> —Roger McNamee, at a Hambrecht and Quist conference, April 1999

Then the world *collapsed.*

The Nasdaq peaked just a smidgen past 5000 after a near-decade run from a mere 333. It then collapsed to 1114—almost an 80 percent drop. The last positive squeal from the space occurred when AOL used its generously valued stock to buy the venerable Time Warner. That squeal lasted a nanosecond. AOL's stock sank from $55 to $11 a share. Other Internet stalwarts were having trouble as well:

Amazon sank from $64 to $19.
Yahoo! sank from $89 to $7.

And these were survivors. Most *didn't* survive.

In 1995, there were eighty tech-related initial public offerings. At the peak in 1999, there were 139. In 2001? Only seventeen. And 2002? Just twelve.

CRISIS? Yea . . .

In September of 2000, when everyone on Wall Street was mulling over their 2001 earnings projections, I remember riding in a car one night in London while talking on a mobile with a favorite client, Bob Rezaee, who was in San Francisco. I told him I thought 2001 aggregate earnings-growth estimates for technology stocks—which seemed optimistic to me at 24 percent—were too high. Given that 24 percent *was*

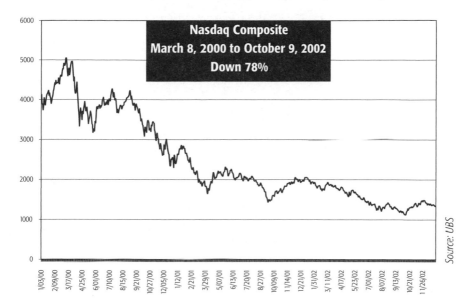

Source: UBS

the consensus at the time, I was clearly in the minority. Bob asked how low I thought the real growth might be—just 10 percent? Or even flat? I said, "Sure," being an open-minded change-oriented person and all, but even I couldn't fathom a decline in earnings. A reduction in earnings? No way.

Technology earnings don't fall. They rise. That's what they do for a living, and the only question is *how fast?* Everyone knows at least that much, so anyone so far out of the box to suggest that earnings could drop would have been offered sedation en route to the psychiatric clinic.

To relieve the suspense . . . were earnings up 24 percent in 2001?
No.

Up 10 percent?
Nope.

Certainly they managed to break even?
Wrong.

So it was a crisis! Earnings fell something like 10 percent, right? That would be quite a shock to a system that had seen nothing but 25 percent annual growth in the 1990s!
That does sound like a crisis, but no, not exactly.

So what happened?
Let's try—down 65 percent from 2000 to 2001!

The investment community expected a 24 percent increase and real life delivered a 65 percent decline!

Expected: +24 percent
Experienced: −65 percent
−65%!

Nothing like this had ever happened in the fifty years of the tech ecosystem. No one would skate through what Microsoft CEO Steve Ballmer called the collapse of three bubbles: the Internet Bubble, the Telecom Bubble, and the Y2K Bubble.

Unit growth of mobile phone sales dropped from plus-50 percent in 2000 to minus-4 percent in 2001.

Unit growth of PC sales went from the midteens to barely positive in a one-year span.

Growth in global corporate IT spending dropped from 12 percent in 2000 to 2 percent in 2001.

Global semiconductor revenues were down 32 percent in 2001, *just a tad (sarcasm)* below the long run plus-17 percent established trend line.

Cisco took a $2.2 billion write-off of its component inventory in April 2001. This is a very important point: what was arguably the

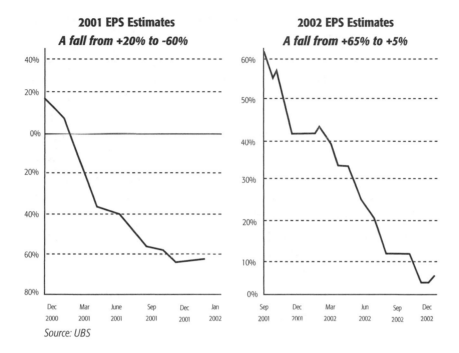

2001 EPS Estimates
A fall from +20% to -60%

2002 EPS Estimates
A fall from +65% to +5%

Source: UBS

smartest technology company during the entire 1990s not only failed to see the downturn right in front of its nose but actually built significant component inventory in the expectation that the 1990s trends would last *many* more years. Securing the necessary components to build the products everyone wanted—while simultaneously keeping those components from competitors—seemed a strategy to help Cisco control its own fate. It was an expensive mistake.

What about Microsoft—the cornerstone of the PC age and the company that nearly missed the Internet until it didn't? Microsoft loved its salad days of the early nineties and its 45 percent compound-revenue growth rate between 1990 and 1996 so much that the company pretended to still be more or less a growth company for a few years after its prime. They bought stakes in AT&T and Comcast to get new devices through those channels and seemed aimed toward the same goal in Europe until the European antitrust authorities stopped it quick. But none of it worked anyway. Microsoft thought it could dominate other corners of our digital lives like it had dominated the PC. None of it worked.

Fortunately for them, Microsoft seems to love winning even more than it loves growth. After bouts of losing in the late nineties while everyone else seemed to be winning, the company decided to change the game. Great idea! Microsoft is on its way to becoming an important and effective *mature* technology company. Today it is a far different story. Microsoft has settled its legal issues with old nemesis Sun Microsystems and many others. It has ended its option expense program in response to investor complaints. It is "solving" the *boring* but critical task of security on PCs. It is targeting a $1 billion reduction in operating expenses. These are all signals of a company that wakes up one day to realize and admit that the world has changed.

In the parlance of this book: Microsoft's crisis in wanting the experience of winning was greater than its perceived pain of changing. So it changed.

Microsoft is accepting, adapting, and changing, And Microsoft is hearing typical Earthlings yelling:

Make technology work!
Make technology easy!
Make technology reliable!
Make technology simple!

CRISIS? Yea . . .

FOUR

WHY TECHNOLOGY FAILS

"Those who do not remember the past are condemned to repeat it."

—George Santayana

"The history of technology predictions is a resource to be mined not a pile of failed futurology to lampoon."

—Michael Shrage, *MIT Technology Review*, October 2004

In 1995, at a twelve-person off site, one of my mentors, Eldon Mayer, picked on one of my recent investment write-ups. He was afraid I had become infatuated with knowing the unknowable. The write-up in question was on the set-top box maker General Instruments, if you remember them at all. They eventually blew up, unable to finalize their digital set-top box technology, which was way too early for the market anyway. They were never able to take advantage of international opportunities. Motorola eventually swallowed them whole.

Eldon feared that the exceptional thoroughness of the document I had created—which included a multiyear financial analysis and an intricate and elaborate valuation schema—indicated that I had been lured by the left-brain sirens into thinking that I was capable of divining the future with precision. The reality, as Eldon pointed out, was that I didn't

THE CHRONOLOGY OF A BROKEN CULTURE

- Technologists interface with other technologists in a "specialized" industry.
- A culture of pursuing deep-seated spec-oriented R & D is set.
- Successes are judged to have sprung from that culture as opposed to in spite of that culture.
- Selective amnesia thrives.

- The PC becomes a mainstream technology.
- The success of the PC is credited to the culture technologists had created.
- But what really happened? A "specialized" industry became mainstream.
- End user requirements had wholly changed.
- Still, the success of the PC reinforces the technologist culture.
- Technologists believe it was mankind's adaptation of itself to the PC that drove the PC's success.

- More supplier-oriented gee-whiz technology creation is encouraged.
- Selective amnesia wears thin as clearly disruptive 10x Grove's-Law changes are no longer commercial successes.
- Crisis ensues.
- Selective amnesia ends.

- A witch hunt to explain failure begins.
- Everyone notices that supreme logistics company Dell is winning without R & D.
- Technologists are blamed and seen as indulgent.

- Technologists lose their voice in the discussion.
- Technologists lose control.
- And the search for new answers begins.

- An examination of past failures is now politically possible.

Hopefully:
- We learn.
- New theories emerge.
- *Change happens.*

know if General Instruments would earn $1.83 per share in two years instead of the consensus $1.75 or whatever the numbers may have been.

Eldon believed that humans aren't capable of knowing the future with the type of detail they desire. However, he also believed that this uncertainty, this lack of detail, could be used to our advantage, rather than serve as our noose. Here was the investment philosophy Eldon passed on to me—figure out what the big changes are, see why we believe in them deeply, see if we think we are the last kids on the block to figure this whole thing out, and then either jump in or not. It was simple, not intricate or elaborate. The answers won't be found on line 262 of a wonderfully intricate 364-line spreadsheet.

One of the most prominent tech investors over the past fifteen years, Roger McNamee, put it this way: "Life happens to the left of the decimal point."

The Change Function is about understanding the big abstracts and elements that are critical in assessing change. Change will not be found in even the most enthusiastic of all spreadsheets or, as a Wharton professor put it to us long ago:

"If you need to calculate a net present value to see if a project is a good idea or not, it probably isn't."

The goal of examining past failures is to learn. And to apply the Change Function to see what we might uncover through a different lens.

The Change Function
f (Perceived crisis vs. Total perceived pain of adoption)

A few last words before jumping right in:

I don't blame the technology for not being ready. Technology "failures" are failures to meet sales and profit expectations. More often than not, the technology itself works, but the user did not care or the business models were flawed. But I don't blame the technologists for not delivering technology.

I don't blame price being too high. An advantage of the Change Function is that we can remove price from consideration by hypothesizing what the crisis on one side would be and what the total pain of adoption would be if price was not an issue. Would anyone care? In most failures, my analysis suggests that the total perceived pain of adoption was a *far* greater influence on the final decision than price itself.

I don't assume the world is static. Cultures, wants, crises, and the perceived pain of adoption all change across time. The mediocre conditions for a technology to be adopted in year X may become quite favorable in year X + 20.

The study of change—and the failure to change—can help us become better entrepreneurs, better managers, and better investors.

> *"It is the mark of an educated mind to expect that amount of exactness which the nature of a particular subject admits. It is equally unreasonable to accept merely probable conclusions from a mathematician and to demand strict demonstration from an orator."*
>
> —Aristotle's *Nichomachean Ethics*, book 1

FAILURE 1: THE PICTUREPHONE

"Some day you'll be a star!"

—Advertising slogan used decades ago by the Bell system to
promote the futuristic communications device called the
Picturephone

Crisis: None
TPPA: High

Background

Although it was developed in 1956 at Bell Labs as a prototype, it wasn't
until the 1964 World's Fair in New York that AT&T publicly launched
the Picturephone—a device consisting of a telephone handset and a
small TV. Shortly thereafter, AT&T announced Picturephone service between New York City, Chicago, and Washington, D.C. The cost: $21 for a
three-minute call—and that was on top of the hardware cost of around
$500,000 for the Picturephone itself. AT&T envisioned this business ser-
vice graduating to the home in time.

Despite survey results indicating most people did not like the Picturephone because the controls were awkward, the picture was small,
and most people were not comfortable with the idea of being seen during a phone conversation, AT&T executives were convinced that it would
find a market and that three million units would be operating in homes
and offices by the mid-1980s, bringing in revenues of $5 billion a year.
None of that happened. Picturephone even failed to find a home in its
initial business target market.

The Upshot

AT&T banked on network effects to drive adoption of its Picturephone.
What it didn't see was that customers had little innate desire to see live fa-

cial expressions on the other end of the line. In other words, there was very little user crisis. On the other side of the Change Function, there was also a high TPPA of waiting around for the Picturephone network to build. It never did, and the Picturephone became one of the most visible flops in communications technology history. Four decades later, today's Picturephones still present innumerable perceived pains of adoption.

Crisis

Was there anything remotely resembling a crisis, or was the Picture-phone a solution looking for a problem? Were there that many businesses in 1964 with heavy phone traffic specifically between New York, Chicago, and Washington, D.C., that was so important that users would feel compelled to lock up Picturephone service between these three cities but no other? Was it even remotely tempting for folks to want Picturephone capability or does this seem like a case of *if we can, we should*—cut from the very same cloth as Microsoft's recently failed SPOT wristwatch. Note to the gadget-heads in Redmond: Just because you can be Dick Tracy doesn't mean you should be Dick Tracy or that anyone else will want to be Dick Tracy!

Microsoft's failure to uncover a crisis even in 2004 is similar to AT&T's failure to realize that although we liked watching *The Jetsons* on TV in the 1960s, no one was foaming at the mouth to actually *be* George Jetson.

In the case of the Picturephone, there was no strong societal reference point for its take-up. As my friend and change guru Zach Karabell suggested back in May 2004, most successful technologies have reference points that society can "get," and those reference points are not extreme. In other words, while technologies may seem disruptive, they are much more likely to be adopted if they offer *incremental* adjustments to new users and not complete deviations from life as we know it. In the case of the Picturephone, AT&T was a generation away from having a prayer. There were no societal references to speak of.

In the 1960s, communication was far more limited than today. In my house, even straight through to the late 1980s, the words *it's long distance* were frequently uttered. For those of you who don't remember the good or bad of the 1980s—Duran Duran perhaps crossing both categories simultaneously—the phrase *it's long distance* meant either *get to the phone quick* or *get off the phone quick,* because the call was bound to be horrifically expensive. Talking to my sister at college in Indiana in 1975 from Rochester, New York, was a *three-minute max* conversation, although my mom often found it in her heart to let four minutes prevail.

To suggest that local area picture-phoning made sense would be to miss the lack of crisis in 1964 to actually see someone who might be just across town. Why in the world would it make sense to use a Picturephone if you could drive twenty minutes to see someone if you really wanted to?

The Picturephone demonstrated that living a little bit like the Jetsons was possible. But the Jetsons were captivating because they were futuristic. *The Jetsons* made a great reference point for TV, not for real life. The future was the reference point of *The Jetsons*—a way of living so far away that it was interesting to watch or contemplate. It was not interesting to actually live that far in the future in the present day.

Total Perceived Pain of Adoption

Did Gordon Moore's Law have a fighting chance of weaving magic with the Picturephone?

Nah!

At risk of alienating those readers already grappling with Moore's and Grove's Laws, I will now introduce another. Called Metcalfe's Law, it states that the value of a network is the square of the number of its participants. The value of a network, the thinking goes, grows exponentially larger the more people that join it. Witness the success of eBay—the sellers go to where the buyers are who go where the sellers are, and so on. What about the fax machine? Who bought the first one and to whom did they fax something? The millionth fax buyer had every incentive to get connected, but the first buyer had none. Metcalfe's Law talks about the power of such networks as they grow.

Metcalfe's Law:
The value of a network is the square of the number of participants.

So let's apply Metcalfe's Law to the Picturephone. To the first customer, the value of adopting a Picturephone would be zero. (Or, technically, the value would be one. One participant, squared, is still one.) For the millionth customer, on the other hand, there would be 999,999 other people to call, and thus the value starts to be both real and large. And thus we come to a common problem in technology that's dependent on the network effect: how do you inspire early investment in infrastructure at the individual level when there appears to be zero immediate benefit? The fax, it should be noted, took twenty to thirty years to click into its network effects.

The network effects are not merely about early *investment*. It's never the money, so they say. The *pervasiveness* that may come along with investment affects the solution set that users consider. The Internet also stood as an available medium for twenty to thirty years prior to clicking. The Internet required modest *financial* investment.

Whatever way ubiquity happens—whether financially demanding or not—it affects our top-of-mind solution set. It is about mind share. Net-

X		X^2
1	$\longrightarrow$	1
10	$\longrightarrow$	100
100		10,000
1,000	$\longrightarrow$	1,000,000
10,000	$\longrightarrow$	100,000,000

Source: UBS

work effects are wonderful when they go your way, but until they do, network effects present a barrier to mind share. An example: At Vortex 2003 we were told that in the United States alone there were nearly 370,000 ATMs. So if you've got an ATM card, you can darn well be confident that you will be able to use it to get money nearly any place you wander with little hunting. With all this convenience, using your ATM card *is* a top-of-mind solution for accessing money—with a low total perceived pain of adoption.

Today, decades after the Picturephone was introduced, even more modern video-related products have yet to become pervasive. Even after the World Trade Center tragedy in September 2001, which people suspected would lead to less travel and inspire video conferencing, widespread adoption hasn't happened.

The Picturephone failed to move the needle an inch. So, too, have today's PC-based personal video-conferencing systems. Few would argue that price has much to do with the present-day failure of video-conferencing to become pervasive. Why is there still a problem?

When I was at UBS, the IT staff installed a PC-based Avistar video phone system. It was an expensive system, and the perceived benefit wasn't fully accepted, so UBS didn't roll it out to everyone at once. That left the company in a chicken and egg situation as far as mind share goes. Here's what happened:

- **Ease of Dial:** Contacting someone on Avistar was quite easy. Well, technically it was, but since I didn't use it much, it clearly did require a mental nudge, but let's score it up as easy.
- **The Hit Rate:** My confidence in actually connecting with someone on Avistar was very low. Or would have been if I'd even considered it as a normal option.
- **Asynchronous, Antimobility:** Most of the people I might have wanted to communicate with over Avistar moved around a lot. Heck, we don't expect folks to actually be at their desks nowadays, so why would we invest the effort to find out that they aren't there anyway? Why not go to the phone straight away with a fallback position of voice mail? Or e-mail? Or chat? Or text

message? Or IM? We have many more choices today than in the 1960s.

- **Asynchronous Life:** I may want asynchronous communication anyway—meaning I don't really want to talk to this person live at all.
- **Alternatives:** Most of my communications—90 percent-plus—occur at times when I am away from my desk. Avistar is not on my mobile devices, while BlackBerry and my cellular phone service most certainly are. Giving away video conferencing on 3G *may*, on the other hand, incite a crisis.
- **Closed System:** Most of my communication was external to UBS. Avistar was a closed internal system, so that limited its usefulness.
- **Camera Shy:** Lots of people don't want to be seen spur of the moment picking up the phone—or even at all. I avoided mentioning this earlier so that the failure of the Picturephone would not be reduced to *just* this—the issues run far, far deeper.
- **A Question of Value:** In how many situations is video an actual enhancement to communication? When my wife wants me to pick up eggs and milk? Probably not so much. When I need to see every nuance of communication? Yep. How often is that?
- **Losing the Mind-Share Game:** Avistar wasn't top-of-mind because it didn't work very often in solving my communication needs.

When did I use Avistar? When communicating with tech-enthusiast and early adopter Sean Debow in Hong Kong after he began working with me at UBS in 2001. These were prearranged meetings on Thursday nights in New York when we worked quite late each week. But we ultimately decided that the phone worked just as well. My boss Alan in London liked conducting my review on Avistar, but more often than not I was on the road at such times or he was in New York live, so Avistar wasn't so necessary. Technology guru and Bell Labs veteran Bob Lucky only chatted with his boss. He remembers those as largely awkward moments when they just stared at each other uncomfortable without protocol for the medium.

Before I move on, I would be remiss if I failed to mention a classic moment in my Avistar life. My colleague Dave Bujnowski used to work in San Francisco a long time ago, and occasionally we would use Avistar with him. On one occasion, I found myself working out of the San Francisco office with Dave and we decide to Avistar with the rest of our group—Diana, Faye, Qi, and Faz—back in New York. Suddenly a real live New York City mouse invaded their conference room, which we couldn't see but *they* could, and Dave and I—thanks to Avistar—watched four grown people desperately climbing for higher and safer ground screaming all the while because of a silly little mouse. Maybe the Picturephone does have some killer apps after all.

Lesson 1

Picturephone: No crisis for users = technology looking for a problem → the number one problem with tech development.

Lesson 2

Futurist dilemma = users don't want to live in the deep future → too great a leap. Futurist visions suffer without an anchor for extension or metaphor from a currently used technology.

Lesson 3

Network effect creation challenges = barrier to user mind share. Want to be a top-of-mind solution!

Lesson 4

The Change Function is dynamic not static. Parallel user problems to video calling still exist → there were core sociological issues with the

Picturephone, some of which have been addressed—but new issues have arisen making teleconferencing still challenged today.

FAILURE 2: INTERACTIVE TV

"Despite the predictions of technofuturists such as Larry Ellison, the companies weren't sure they would ever see a return on their investment. Would large numbers of people actually subscribe to a video on demand service? Or would they just keep making the trip to the neighborhood video store? Nobody really knew. And nobody wanted to bet a few hundred million dollars on the answer."

—Mike Wilson, *The Difference Between God and Larry Ellison*

Crisis: Very Low
TPPA: High

Background

Interactive TV (ITV)—programming that incorporates interactive graphical "enhancements" such as icons, banners, menus, text fields, streaming video, and Web pages—generated massive buzz in the mid-1990s. It was heralded as a revolution in the way viewers would watch TV—among other things, they would be able to instantly read more about the topic presented during a program, download and store related media files, purchase goods, conduct banking activities, and share in real time their knowledge or views about the broadcast. At the time, Myers Reports projected that ITV would reach annual revenues of more than $32 billion by 2006.

In an effort to capitalize on these rosy projections, many major U.S. companies ventured into the realm of ITV, including AOLTV, Bell Atlantic, MTV, Oracle, TBS, and Time Warner Cable in New York City. But Interactive TV faced many challenges to adoption.

Ignoring the technological challenges inherent in producing and providing actual ITV content, the failure of the various corporate players to agree on a worldwide standard meant that ITV content was also extremely expensive to provide. Financing ITV was difficult because subscribers weren't willing to pay prices that would cover the advanced services costs, and the targeted advertising programs where ITV providers recovered the remaining balance of the cost were hotly debated as potential invasions of privacy. Due to these challenges, mid-1990s ITV failed to meet its promise and was pronounced *dead*.

The Upshot

Interactive TV promised neat things: video on demand, buying stuff on the Home Shopping Network on a remote control instead of a cordless phone. But the advantages that Interactive TV offered over existing processes were small. Meanwhile, installing Interactive TV and learning how to use a five hundred-button remote control presented enormous learning curves—in other words, a high TPPA—that customers were unwilling to tackle when the status quo worked just fine. The hype died when shareholders rebelled against the huge price tag of the rollout.

Crisis

In *The Difference Between God and Larry Ellison*, Mike Wilson chronicles Larry Ellison's part in hyping Interactive TV—Ellison was one of a gazillion—during the early and mid-1990s. After the death of the Interactive TV hallucination, one might expect Ellison to be embarrassed for being so totally and exquisitely wrong—or at least humbled by the experience. Not the case. Wilson notes:

> *"Why be embarrassed anyway? The hullabaloo over video on demand had landed Ellison on the covers of* Fortune *and* BusinessWeek *magazines; both articles portrayed him as a pretender to the throne of Bill Gates. More important, Oracle had become, if not a household name, at least a recognizable one."*

Fortune reported that, in 1995, 40 percent of respondents recognized the Oracle name—up from a mere 2 percent in 1992.

So who exactly had the crisis? Users? It doesn't seem so—at least back in 1992 or '93 or '94 or '95—back then, *Interactive TV* really meant only a couple of things, the first of which was video on demand. But unlike the success that DVD-rental outfit Netflix has enjoyed in the early portion of the current decade, it wasn't clear at all in the early nineties that folks considered the mildly annoying journey to the video store a *crisis*. What's more, watching a video doesn't seem to qualify as very interactive, but why quibble over semantics?

Interactive TV also meant buying things, and people were indeed buying things on the Home Shopping Network and the like. But the current system worked and had the convenience of employing the low-tech device known as the telephone.

Crisis? Nah . . .

Interactive TV didn't even come close to meeting Andy Grove's suggestion that a new technology deliver a tenfold improvement in performance or capability. By that standard alone, Interactive TV was a flop.

Technologically, Interactive TV was an impressive concept to engineers, but in the 1992 lives of real Earthlings, it promised nothing but pain:

> *So . . . let's see . . . you have a substitute service number 1 to save me time in going to rent a video, which will move that intense ever-present argument of "what are we gonna watch tonight" from the video store back to our living room and then a substitute service number 2 so I can be confused by your remote control when my telephone seems to work pretty well already. No thank you.*

For the slew of other supposed services that were promised—*Hey, I might like buying my pizza with my remote because the telephone is waaaayyyyy over there or I would love to have all of Michael Jordan's high school basketball statistics during the Olympics*—the motto might well have been *the dummies are out there, the applications will come* or *build it, and they will sedate.*

Andy Grove might be embarrassed to hear all this called disruptive, even if Larry Ellison wasn't embarrassed to have captured greater name recognition through the process.

So, again: who had the crisis in the early 1990s? Larry Ellison? Maybe—a crisis in needing new outlets to sell his server software and all those "Interactive" movies would need to be held on servers.

Other technologists? You bet. Technologists yearn to be part of something special in the same way crowds worldwide wanted to believe that Muhammad Ali was The Greatest—because it meant that watching the *Thrilla in Manila* was *really* special. Interactive TV sounded really special. There's nothing wrong with technologists wanting to be a part of history—thank heavens for it.

How about Bell Atlantic CEO Ray Smith, just a few years away from retirement and with the crafting of his legacy on his mind? Could that have made him susceptible to buying into a dream that was still technologically a hallucination with zero crisis for his customers in suburban Philadelphia? What was he feeling while hanging out on renegade John Malone's boat in Boothbay Harbor, Maine, crafting a merger with cable legend TCI to take his boring, dormant, bureaucratic Bell Atlantic into the new world before handing over the reins? Did Ray Smith have a crisis? Maybe.

Did Bell Atlantic shareholders have a crisis? Many of them decided they soon would if Ray Smith's "vision" forced a cut in the reliable dividend emanating from their boring company. They bought shares in Bell Atlantic *because* it was boring and dependable. And they liked it that way! They were wondering if Ray Smith was actually crazy enough to sink the company on a decade-long flier.

Mind you, this was back in a time long before VoIP or other current crises facing the telecom industry, so Bell Atlantic's profit stream from its voice business did seem pretty safe. It was also well before Bell Atlantic would recognize a more blaring and brutally clear crisis— America Online and access to the Internet.

So there was crisis—but not at the end user and not among share-holders of the service providers. The crisis was among those wanting to be visionaries and at the vendor level.

Lesson 79 here:

If the vast bulk of the conversation is *from and about* the purveyors of the new technology, watch out. Of course they *want* to be positive, and in many cases the individuals cited *will* be fired if they don't keep up pretenses. If the remainder of the conversation is from service providers—watch out as well. If there seems to be no discussion from *real, live* users, and if the "trials" seem to be as secretive or covert as the Watergate investigation, then watch out.

The good news is that system worked—not the Interactive TV system but rather the shareholder system. They yelled enough that Ray Smith was unable to complete the well-intentioned but potentially disastrous merger with TCI. That disaster happened on Mike Armstrong's watch at AT&T instead. The trials were summarily canceled around the planet, and Interactive TV was pronounced dead. There had been some cool tech talk by tech folk but never a crisis at the end user level.

Was the pronouncement of death premature?
You bet.

Huh? You just spent three pages laboring on about how there was no crisis about Interactive TV and now you leave the door open. That makes no sense!

Yes and no. There was no crisis in the early 1990s. Will there be a greater crisis as society shifts its sensitivity to what crisis looks like? Sure.

I wasn't alive in the early portion of the twentieth century, but I know the word temperance, so here's an analogy. Apparently in the United States there was such a significant crisis that society went to all the trouble of outlawing alcohol.

Banished . . .

Banished . . .

There must have been a big crisis of some sort to go that far. Fourteen years later, the *very same* society had a new, fresh, and equally powerful crisis and decided to go through all the trouble to *un*outlaw the very thing that had been outlawed.

Crisis shifts.

Crisis is not absolute, Crisis is not always rational. Crisis is a state of mind.

Ten years later, Netflix has a decent business serving customers who would rather ship DVDs through the mail than visit a store. That's their choice—and their crisis. Will society one day think making a trip to the local video store is the dumbest thing one could ever imagine and that they would freak if they had to use phone as opposed to clicking BUY when they wanted that cubic zirconium on home shopping? I wouldn't be surprised.

Total Perceived Pain of Adoption

Do you ever wonder why each successive version of Windows looks pretty much the same as the prior release even though the best way to overcome industry criticism that Microsoft is not "innovative" would be to chuck out the old look in favor of a new one? Ever wonder why Internet Explorer looked pretty much like the already successful Netscape Navigator? Do you think Microsoft will model its own search technology after Google's?

> *"Gates and Allen started Microsoft with a stated mission of putting 'a computer on every desk and in every home, running Microsoft software.' Although it seemed ludicrous at the time, they meant it."*
>
> —Robert Cringely, *Accidental Empires*

Microsoft, when at its best—when it's not creating useless features or allowing security gaps of monumental proportions—is really good at making things easy for normal Earthlings. Bill Gates and Paul Allen had a vision of a computer in every household. They didn't help create and profit from that trend by reacting to every criticism that the technology community had of their company. They didn't try to be popular with techies in search of a piece of the cult status the far poorer Steve Jobs and Apple gained—even as Apple's destiny appeared locked in third-tier status and Microsoft was even bailing them out.

When people claim Microsoft's interfaces are not on the technological leading edge, well, of course they're not. Why force users to keep up with whimsical engineers as opposed to merely serving your customers and making their lives "better" while you keep it simple and easy and not scary? That was a rhetorical question.

Interactive TV was an invitation to shift user interfaces. It was an invitation to abandon the very familiar interface known as the telephone for the complexity of yet another remote control with *way* too many buttons to possibly "get," though it would undoubtedly be logical in the way E=mc^2 was logical to Einstein. This invitation to *come into the new world*—a generator of both guilt and insecurity—was met with *No thank you. . . . I will remain in the old world if you are suggesting I need to become more dependent on that dastardly confusing remote control to do the same cubic zirconium buying that I have always done."*

Technology provider with crisis: "But it's better!"

Earthling without crisis: "How?"

Technology provider with crisis: "It makes more sense."

Earthling without crisis: "You don't listen. There are ninety-six buttons on my remote and I only use fourteen of them because I don't want to learn the others, which I am sure you will say are 'better' and 'make more sense' as well. I have no desire to spend my leisure time feeling stupid just so you will think more highly of me. Your remote is already too complex."

Technology provider with crisis: "But the interface on the TV will allow you to do so much more with your television than you could ever do before. An interface will pop up on the screen and you can sit in your couch while you order your pizza from one of four local chains including Pizza Hut."

Earthling without crisis: "Another bunch of stuff to learn in order to order my pizza from Pizza Hut? I don't even like Pizza Hut. I just want to sit here and change channels in a dull stupor recovering from the world and you want to tax my brain and change my pizza vendor because you both want a slice of my pie. How is that better? My remote control already scares me and you're just looking to sell me junk I don't even need. Now be quiet—I need to buy another cubic zirconium!"

In his recent book about the brain *On Intelligence*, Jeff Hawkins—the developer of the PDA—suggested one thing that struck us as off-base: "To this day I still hear people claim that computers should adapt to users. This isn't always true. Our brains prefer systems that are consistent and predictable and we like learning new skills."

I rarely disagree with Jeff but here I do. I don't think most people like learning. Or perhaps more accurately, I suspect most folks experience learning as so painful that the "crisis" better darned well be high and the benefit both large and immediate. I think people like *having* learned. We continue as a society to actively do things we know are *not* in our individual or collective best interest because doing otherwise demands change from the lifestyles we have gotten used to. I know that sugar is bad for me, but just leave me alone while I finish my bowl of Captain Crunch cereal. People resist learning and changing.

Hawkins's PDA included an innovative pen-based methodology that challenged several million prior to its abandonment. If you don't conform to humans, you may raise the users' *total* perceived pain of adoption to the point that any actual crisis doesn't seem so, well, critical. Separately, couch potatoes want to veg out—the number of people who "like" learning how to use a remote control is exactly forty-two.

> *"One of the earliest innovations to have an influence on the generation that would invent ITV was a show called* Winky Dink—*a program first broadcast in October of 1953 in black and white—on the CBS network.* Winky Dink *featured the adventures of a cartoon character named Winky Dink and his dog Woofer. During the program,* Winky Dink *went on dangerous cartoon adventures and got into a lot of trouble. In order to save him from his perils, Barry came up with a unique gimmick: The Winky Dink Kit. Inside the kit, there were sheets of transparent plastic and several crayons. When prompted on the show, kids would place the plastic on the TV screen and draw a bridge or rope across a cavern or river, as examples, from which Winky Dink could escape."*
>
> —Tracy Swedlow, "2000: Interactive Enhanced Television: A Historical and Critical Perspective"

Lesson 1

Techies *always* have a crisis to sell new technology—so we can't listen to vendors as they describe what the "user" wants—they largely shouldn't be expected to escape their own fishbowl of thinking.

Lesson 2

While every constituent in the technology food chain is a "user" in some way to someone else, the end users are typically the most important.

Lesson 3

Crisis is a dynamic state of mind. Crisis is not absolute. It changes.

Lesson 4

Providing a new user interface based on the already despised remote control? Yikes!

Lesson 5

A technological 10x shift doesn't necessarily make for a 10x enhancement of user experience.

FAILURE 3: IRIDIUM/GLOBALSTAR

> *"We all want to have a communicator like Captain Kirk had so we can talk to anyone anywhere on this planet—at anytime. The real challenge is for the private sector to meet our demand—and adapt to it as our demand changes. Imposing a solution that doesn't meet actual demand (or the imminent prospect thereof) is doomed to failure. As far as Iridium is concerned, the rocket science worked like a charm—It's the underlying economics that failed."*
>
> —Keith Cowing, "Iridium: If You Build It
> They Won't Necessarily Come"

Crisis: None
TPPA: High

Background

Iridium and Globalstar were designed to offer continuous global communications coverage. Motorola's Iridium was a system of seventy-seven satellites that began development in 1990. By 1996, Motorola had raised $1.9 billion of the necessary $5 billion in equipment costs. Although Iridium service debuted in November 1998, problems arose immediately with the phones because they were extremely large and

cumbersome, they were late to ship, they didn't get service inside buildings or without direct line of sight with a satellite, and they weren't capable of handling the high demand for Internet access. The prices
associated with Iridium were also a major impediment to its adoption,
with the phone itself costing around $3,000 and call rates costing
around $7 per minute. By 1999, Iridium had declared bankruptcy, se-
curing a place as one of the most expensive failures in history.

Globalstar, which was established in 1991 by Loral and Qualcomm,
was a direct competitor to Iridium. Its own system of forty-eight Low
Earth Orbiting satellites (LEOs) was designed to provide wireless telephone and telecommunications services including Internet access, position location, SMS (short message service), and call forwarding. The
satellite-based phone system cost $3.8 billion, and first offered commer-
cial service in 1999. Globalstar was significantly cheaper than Iridium—
the phones retailed around $1,500—and it also supported higher data
rates. Despite these improvements, Globalstar was also unsuccessful. On
February 15, 2002, the company closed its doors, leaving its extensive
system of satellites in orbit to burn away over the next ten years.

The concept of a global telecommunications capability for consumers lingers. Craig McCaw followed up with a $10 billion global tele-
com system called Teledesic, with financial backing from Bill Gates,
among others. Teledesic would deliver high-speed video, data, and telephone to anyone, anywhere on Earth. It never got off the ground.

These failures are particularly notable in their seemingly strange
combination of zero crisis and massive funding requirements, coupled
with significant vendor backing and a willfulness on the part of perceived
smart money to participate, even in the face of initial widespread skepticism. Aren't most of these elements usually associated with brilliance,
revolution, and success? Not in the case of Iridium and Globalstar.

If smart moneymakers like legendary Bernard Schwartz of Loral
were not involved, the speed toward failure would have been much
slower. But while faster speed to failure is great for the technology
ecosystem, it was anything but positive for those who coughed up $9
billion to construct Globalstar and Iridium. While the late nineties were
filled with debate over the merits of Globalstar versus Iridium, the con-

DUMB THINGS THAT VERY SMART PEOPLE SAID

"Six hundred forty thousand bytes of memory ought to be enough for anybody."

—Bill Gates, 1981

"I think there is a world market for maybe five computers."

—IBM Chairman Thomas Watson, 1943

"There is no reason for individuals to have a computer in their home."

—Ken Olsen, 1977

"The Internet will catastrophically collapse in 1996."

—Robert Metcalfe

"It would appear that we have reached the limits of what is possible to achieve with computer technology, although one should be careful with such statements, as they tend to sound pretty silly in five years."

—John von Neumann, 1949

versation would have worked better if we had maintained focus on why both ideas would turn out horrendously. The core assumption of demand was *very* faulty, and maybe no one in the inner posse felt much like mentioning it to Schwartz or Chris Galvin at Motorola. Even worse, dissenting voices were likely silenced once the decision to start raising massive amounts of money and the launching of satellites had begun.

There are two key points to focus on here. The first is the concept of suppressed communication. With "big projects," the politics of decisions already made tends to suppress free flow of information until it is too late. This is an example of behavioralism over rationality. Then there's the phenomenon of *ad hominem* in reverse. A large portion of the tech ecosystem will acquiesce amid confusion to a superstar following a new venture. This is the reverse of the illogical *ad hominem* argument. An *ad hominem* argument is one that follows this supposition: If "Jim" thought of the idea, it is a bad one, because Jim is "bad." Here, if Bernard Schwartz thinks it is a good idea, it must be, because Bernard is "good."

The Upshot

Iridium and Globalstar promised that if users were willing to carry around abnormally large cell phones with monstrous antennae, mobile coverage could be 100 percent complete, anytime, anywhere. Did users in the late 1990s want this service so badly—i.e., was it a crisis?—given the inconveniences it required: a huge phone and huge fees? Nah. As I see it, the idea that customers would want to use Iridium or Globalstar—even as a stop-gap in their normal coverage—at the exorbitant fees these services required was ludicrous.

Crisis

What was the crisis? Why did these wonders of technology fail as wireless service providers? Can we count the ways or at least classify them? As with Interactive TV, there seems to be a distinct lack of end user crisis. But someone had a crisis—including those investors who plunked down $9 billion.

So let's start there. In 1999, a good friend, the investor Anthony Xuereb, pointed out that there had never been such a period in time in which every reasonably good idea could be financed. We were in the Golden Age of technology—at the very least, share prices certainly reflected as much. Any difficulty raising money vanishes in such states of investment mania. Too much capital was chasing too few legitimate prospects. The crisis to be part of the phenomenon was extreme, especially for investors. They did not want to miss anything *big* and Iridium and Globalstar were *big*. The crisis spurred investors to jump in while grossly overlooking whether the investments made sense or not. Caught up in the frenzy, they forgot to ask themselves: *Is there a demand, is there a business model?* Investors had a crisis: they feared missing out.

Was Bernard Schwartz's crisis also a fear of missing out? I have no idea, but I can imagine that even the smartest folks on the planet don't want to miss out on the next big thing.

What about Qualcomm and Motorola's crises? Each of these companies also provided "gear" for these systems, and Qualcomm in particular had little to lose. Qualcomm was providing technology—its unique brand of so-called CDMA technology, or code division multiple access, to the process. Qualcomm was having trouble taking its technology and planting it anywhere at the time. And here they suddenly had hundreds of millions of dollars to build out their technology and develop technological reference proofs that CDMA worked in the harshest of situations. Globalstar was a cash cow *and* a stepping-stone for Qualcomm. If Qualcomm had just been about Globalstar, it never would have become one of the twenty most highly valued companies on the planet. But did Qualcomm have expertise in understanding demand for Globalstar?

Not at all.

If we were to look at Qualcomm's track record of investing in the build out of new systems around the planet and, in this case, *around* the planet with LEOs, it would look like a disaster. This seems to suggest that the smart money at Qualcomm is very dumb. But if we consider that Qualcomm used their many investments and their Globalstar relationship to build a business around their CDMA technology, they are utterly brilliant.

What about the end users? Here's what struck me first: The Globalstar marketing material pictured someone standing in the middle of a desert or in the mountains of Tibet or—you get the idea—with the suggestion that they could be connected to anyone on the planet. Pretty darn cool, that, but what's the market for such an application? Next time I'm in the mountains of Tibet, I'll run a survey: if I find another soul, I'll ask. Mind you—and as you probably already know—I'm among the fringe 1 percent or so of people who imagine technology connecting the world in ways John Lennon never could have imagined. But even I balk when I see a marketing brochure with a person standing in a desert making a phone call. We need to balance our wonderment at technology's role in creating nirvana with a skepticism about business models.

So if extremely remote people can't be considered a market, what about the neighboring market of remote villages? If we focus on the possibility that phones would allow a way to communicate with distributors of farm products or to receive remote medical help and the like, Globalstar was still talking about selling a single phone to a village whose villagers would then line up to share it. How many of those villages are there? More importantly, how would you go about establishing a distribution network for selling to these villages? Talk about high costs of penetrating a potential client base.

Were telecommunications service providers so desperate for fresh revenues that they would partner with an Iridium or Globalstar to perform such a function?

Nope.

Globalstar's plan, which was dependent upon incumbent service providers creating the distribution channel, was dubious. Bernard Schwartz figured that if the system made enough sense, he would be able to tie it back to an economic play for an array of the world's most prominent service providers. Certainly the long list of named partners was impressive. Vodafone and AirTouch were two anchors. But the word *partner* doesn't mean much if it's only on a press release. These were *partners* in the minimum sense that their names appeared on a slide that Globalstar presented at a slew of large investor gatherings known as rubber-chicken lunches.

> *"If all these fine household names let us put their names on our slides, it means that they don't think we're insane, because they wouldn't let us do that if they did, right?"*
> "No comment."

Was there an economic incentive for service providers to act as distributors? Not really—mainly because of scale. Globalstar's optimistic projections were a drop in the bucket for its distribution partners. Why would they put a lot of resources into making this complex scheme work? They never did.

> *"Okay, okay, okay—but you were just kidding that connecting the rural villages of the world would provide a market itself for LEO voice systems. It was all about 'infill,' wasn't it?"*
>
> "Yes, Iridium and Globalstar were meant to be global 'infill' systems for the basic mobile phone. In other words, where there were gaps in coverage when I drove up the Merritt Parkway in Connecticut or through "rural" Chappaqua to the Westchester airport north of New York City, I would turn off Verizon and pull out my monstrous Globalstar phone with the antenna that reaches the heavens and chitchat at outrageous fees for five minutes before turning it off to go back to Verizon once digital coverage resumed."

For business people heading to Europe in 1995, the idea of global mobile coverage was mainly just *an idea*. No one really considered lugging around a bulky Iridium phone while visiting with clients in Munich. The Iridium phone would probably be just as useful on a weekend getaway in the Adirondacks.

Here was the key issue: both Iridium and Globalstar proposed to solve a problem we'd already become inured to. By 1997, almost all of us had come to expect bad coverage—indeed, sometimes no coverage at all—as a sacrifice for mobility. We had been trained to marvel at the miracle of mobile voice. For me, at least, the "dropped call" became a unifying global experience that I could draw on to connect with any investors I might meet with, anywhere on the planet.

So we might yell and scream in our cars when suffering a dropped call on the way to the airport—but we tolerate it.

Broken mobile phone? Crisis

Short battery life? Crisis

Limited ring tones? Crisis

Unable to place a call while touring Greenland? No crisis.

Total Perceived Pain of Adoption

Iridium and Globalstar each felt that they could charge a substantive premium for infill service that held only modest utility beyond basic mobile service itself. There were many problems with their thinking.

First, their projections were *way* off: Globalstar figured that the price of basic mobile service would fall only modestly and that the premiums for this infill service would be comparably tolerable. Wrong. Prices for basic service have plummeted over the years and minutes of use have proven quite sensitive to price reductions. In other words, price was a big part of the overall total perceived pain of using more minutes for users, and Globalstar didn't adequately anticipate that. I suggested earlier that Moore's Law was a necessary though not a sufficient condition for technology adoption. In this case, Globalstar was going to start charging a huge premium to a service Moore's Law had already helped become somewhat "affordable."

A second problem? How about an additional phone to haul around? Yuck. A third? The phone looked like a brick. For a couple of years when traveling in Asia, I suffered Sean Debow's abuse when carrying a UBS-standard-issue Nextel global phone that was twice the size of Sean's latest gadget. The size demands were growing obvious, but Globalstar and Iridium overlooked the user when lured by the siren's voice to get the technology working at all costs. A fourth? The oversized antenna that was required in order to reach a satellite as opposed to a cell tower.

One last one for good measure? Time. The development of smaller sized phones with normal sized or even internal antennas that easily provide global coverage via seamless distribution agreements eliminated the supposed market for Globalstar and Iridium. The infill the LEO guys suggested they would provide was already being provided by others—much more cheaply so—fairly shortly after all the rubber-chicken lunches.

But all of this high total perceived pain of adoption obscures the more relevant point: the user crisis was so small that even a reasonable TPPA would not have done the trick.

> *"One of the unexpected benefits of the Iridium project was catching 'Iridium Flares.' As the satellite passes overhead and passes into the right configuration with respect to the sun and a viewer on the ground, a brief bright flash of light occurs. These flashes last only a few seconds but can be brighter than the planet Venus. It is ironic that the only lasting visible legacy of Iridium over the next few years will be to serve as flashes in the sky as they orbit overhead and then crash back to Earth."*
>
> —Keith Cowing, "Iridium: If You Build It They Won't Necessarily Come" (spacenet.com)

Lesson 1

"Smart" people do *not* always know what the right answer is when it comes to technology adoption!

Lesson 2

Communication is suppressed once past a point of no return—don't expect key players to fess up that their plan is failing miserably.

Lesson 3

Banking on indirect distribution of miracles is a big risk.

miracle technology does not remove the need to effectively model the business.

Lesson 5

Today's miracle technology may be eclipsed in a brutally short time by a new technology solution for the same problem.

FAILURE 4: TABLET PCS

"Gates may be proved correct eventually, but, two years later, Tablet PC sales remain disappointing."

—Anne Chen, *eWeek*, May 24, 2004

Crisis: None
TPPA: Huge

Background

When Bill Gates announced the launch of the Tablet PC at his 2001 Comdex keynote appearance, he referred to the device as "an evolution of the laptop." Tablet PCs are portable computers, similar to laptops, fitted with pens, touch screens, and handwriting-recognition technology that were developed for on-the-go ink-enabled applications. The Tablet PC can be docked to function as a desktop, attached to a portable keyboard to function as a traditional laptop, or operate through handwriting recognition. There are two basic styles of Tablet PCs—one is a laptop with a screen that can be twisted around backward and then folded down to create a tablet and the other is a pure tablet that resembles an Etch-a-Sketch. The Tablet PC promised to increase daily productivity, save time, and allow the creative process to be more focused.

Faced with PC-market crash in 2001, many top name PC vendors quickly entered the Tablet PC market including Toshiba, Acer, Fujitsu, ViewSonic, Motion Computing, and Hewlett-Packard. Although these vendors had high hopes that a significant portion of corporate laptop buyers would choose a Tablet PC, the market experienced mediocre sales from the start. Tablet PC unit sales were still only around 500,000 in 2002—half of what Microsoft had expected. Acer's Tablet PC sales currently account for less than 10 percent of their total notebook sales and HP's account for just 2 or 3 percent of notebook shipments.

Full disclosure: my wife, Kelly, and our former au pair Ivan each bought a Tablet PC in 2003. Once purchased, these products were never shared with me or our three children—Bailey, Tucker, and Eamon—contrary to our family ethos of sharing everything. Kelly diagrammed yoga poses for her course work, but no one else got to even play with this gee-whiz technology. I could be carrying around some pent-up bitterness from the whole episode, but I don't think so. Two years later, Kelly has abandoned any use of her Tablet PC.

The Upshot

The masterminds behind the Tablet PC underestimated people's fear of change. Few of us desire to learn a new way of using a device when the status quo works well enough. The Tablet PC folks also highly overestimated the crisis in the market. Were people ready to ditch their laptops or pen and paper altogether to embrace the possibilities provided by the Tablet PC? No. Few people today would consider going back to writing a tenfold improvement on hunt and peck. In sum, while the technology behind the Tablet PC was very impressive, the Tablet PC itself was a dramatic step backward for actual user experience.

Crisis

Let's define crisis through a series of questions:

"Would PC users have a crisis if they needed to replace a broken PC?"
"Sure. The PC is among those items that need to be replaced immediately upon their demise, just like a mobile phone."

"Would users have a crisis to demand mobility in their PC solution?"
"Many would say yes, and therefore a Tablet PC qualifies under the broader heading of notebooks."

"Would users have a crisis if their next PC purchase did not have hand recognition capability so that they could take notes by hand or draw pictures or have a rotating screen?"
"No. There's certainly no crisis for most of us here in 2005, although my wife was temporarily hooked on these features."

So, there's not much of a handwriting recognition crisis today. That's a problem. But let's take a look at how various crises built in my own home.

Crisis 1: Our twenty-one-year-old au pair from the Czech Republic, Ivan, is obsessive about using technology, and inordinately comfortable with virtually any machine on the planet. He thoroughly enjoys the arduous process of learning new technologies. Ivan would often interface with the UBS IT folks on my behalf just when the conversation was going to get truly technical and likely last for hours. For example, he helped to troubleshoot my home router. What a guy!

Ivan's crisis when buying a new laptop would have been in *not* getting the Tablet. Once he got the Tablet, he was so excited that despite being a Digital Native—those who are more technologically adept than us older folk, since they actually grew up with computers—and consequently, someone who never has to read instruction manuals—he decided to read this one anyway. This, of course, classifies Ivan as an obsessive-compulsive technology enthusiast in addition to being a Digital Native. It's good to have one of them around.

Crisis 2: Kelly wants to be able to use her laptop for her work; we want a family laptop to use when at a second residence in Maine; and Ivan is the conduit for a purchase. Still living in the glow of his magnificent Tablet experience, Ivan raises it as one of several possibilities for Kelly and—the horses are off and running—she, upon hearing that she can draw pictures that will be saved to memory, is sold on the idea. Ivan will wind up selling four or five more Tablet PCs over the following three months, generating no commissions but lots of appreciation and self-satisfaction.

Crisis 3: The Tablet arrives at our home at 10 A.M. one day. By the time I arrive home that evening, no one else is allowed to use the supposed "family" laptop without Kelly's consent. Kelly—having a great teacher in Ivan—adores her new Tablet. Tasha the cat is feeling displaced. The new crisis: we still don't have a family laptop.

What's really been bought here? Ivan has bought a continuation of his bleeding-edge lifestyle. Kelly . . . well, Kelly really bought an application or a feature as opposed to a Tablet PC.

"Huh?"

She bought a feature—she wanted to diagram yoga poses. She was not abandoning the QWERTY keyboard in any way, shape, or form. I suspect that the problem for the Tablet PC is that *very* few people are seeking a new way to interface with a computer. I also suspect very few people are in crisis when it comes to their keyboards. While we are raising a society of carpal tunnel victims as a reward for our keyboard mania, we are no longer a society of handwriting enthusiasts. Few think that even the advanced handwriting recognition capability Microsoft offers makes the notebook experience notably better.

Most people who buy the Tablet are buying it for specific applications or for a specific feature. My own eventual Tablet purchase would be a function of an application: I like to write and diagram upside down in meetings so clients can see my arguments facing them, thereby hopefully reducing their total perceived pain of adoption of hearing yet another person's perspective on technology investing. In my case, the Tablet becomes a replacement for paper and markers. But how many such diagramists are out there?

How could Microsoft, the company that more than any brought personal computing to the masses, be sucked into thinking that the Tablet PC and handwriting recognition technology was actually an answer to a widespread crisis? Well, for starters maybe Microsoft *isn't* perfectly in sync with the common man's technology crises. As evidence, I note Microsoft's historical obsession with crafting more and more useless features. Rumors often swirl of studies suggesting that less than 10 percent—or even some low percentage like .005 percent—of features are ever used at all.

So when folks refer to the "user-friendly Microsoft culture"—well, they don't actually do it that often, but suppose they were to confuse Microsoft's dominance with a user-friendly culture—it might be a misinterpretation. Microsoft isn't on the vanguard of user experience.

"But Microsoft is smart, isn't it?"

Smart companies take a flier on lots of occasions. And Microsoft is a smart company. In the late 1990s Microsoft invested $1 billion in cable giant Comcast and another $5 billion in then cable giant AT&T. Many suspected that Microsoft was aiming to control the distribution of set-top boxes in order to favor its own Windows CE operating system in the event that set-tops became a larger portion of the big picture of computing in the household.

Did Microsoft know or suspect that the cable guys would win the battle for the living room? Nah. Would Microsoft as a smart, powerful, resource-filled, and resourceful monopolist think to protect all flanks? You betcha. And does Microsoft want to sell as many licenses as possible to each of us? Yep. Does the development of the Xbox qualify as well? A flier that ended up working? Absolutely. Could Microsoft ultimately push the Tablet PC into crisis play? Sure, with a little bit of persistence.

The marketing could feature ways and places in which a Tablet PC simply works *way* better than the alternatives—Andy Grove 10x disruptively better—in order to complement its cool "show-and-tell" image. Students might be a good group to target. They might also look at folks who need to write like, geez, maybe lawyers?

Or perhaps they can target situations where writing in a notebook is far more politically correct than banging on a keyboard. Most of my own meetings during the past ten years have worked better with casual handwritten notes. The use of a keyboard tends to terrify people who grow wary of our interaction as they wonder what is on the other side of our Battleship screen.

Total Perceived Pain of Adoption

Maybe Microsoft's enthusiasm for handwriting recognition technology is just a few years ahead of the curve. For now, the Tablet PC is merely a reminder of what we hated about the Newton and the Palm—the Tablet demands learning for most people. Kelly will tell you that learning how to use a Tablet was both doable and worth it—but she learned under the supreme and patient tutelage of Ivan, the obsessive technology enthusiast.

So for Kelly—a yoga instructor with a need to draw and store diagrams—the TPPA was below her crisis level. But since the crisis for most is quite low, the TPPA must be brought down significantly even in a price-neutral world.

The introduction of a pen rekindles the nightmare from the introduction of the PDA, even if in this case you are *not* asked to learn a new alphabet. Still in the case of the Tablet, the introduction of a third interface—to "complement" the keyboard and the mouse—requires learning for any payback; and as I suggested earlier, Earthlings do not like to learn.

Earthlings like *having* learned but rarely enjoy the *process* of learning. Great teachers are able to help make the learning process less scary. These teachers are the number one facilitator of new thinking. Unless Microsoft can become that "great teacher," we have a society of folks wondering if learning to use this new technology is worth the struggle.

The total perceived pain of adoption portion of the Change Function includes the word *perceived* in order to highlight that one's view of "pain" is personal and ultimately a perception. If users associate a Tablet PC with a pen-based PDA, there is a problem.

The Tablet PC's value proposition as it stands: for more money than you would pay for a regular laptop, feel free to spend hours a day *trying* to understand this new machine you bought—a machine that *maybe* one day you'll decide to find a real reason to utilize once the show-and-tell-your-friends portion of the experience runs thin. After all, twisting and contorting the screen only amuses for so long.

Marketing might be best used to clarify that little is required to learn this new technology—that just because there is writing involved does not mean there exists a new language to learn. Of course, it's always best if there is *zero* learning involved in a new technology. At its core, hand-writing recognition is meant to adapt to its users not vice versa. The message Microsoft should be aiming to get across: it can be *very* easy to learn how to use a Tablet.

Another marketing tack: Microsoft should hammer away at how much it has improved the quality of the handwriting recognition so that you—the user—no longer have to conform to the machine because the machine conforms to you! Unfortunately, most folks either think or know that the machine still demands that you conform to *it*, as opposed to the other way around—a big perceived hurdle to overcome.

> *"There are no miracles. I got a C in handwriting in grade school. Having used the tablet, I understand why I got a C."*
>
> —John McKinley, global technology head, Merrill Lynch,
> quoted on News.com, November 6, 2002

People have been enjoying the uniqueness of their handwriting for-ever. Remember the passive resistant nature of not mimicking our kindergarten teachers when they told us exactly how letters were sup-posed to look? We didn't fail in life as a result. Then we tiptoed out a bit further to develop our own handwriting style and to admire the illegible scrawl of successful physicians around the globe who also apparently did okay without conforming to their kindergarten teachers' demands. As a society, we seek out celebrity autographs for their uniqueness. Any reminder to conform or limit our handwriting will be met with resis-tance.

Such is the fear that the Tablet is saying *Hey, if you promise to be a good little boy or girl and write like we say and pay us a few hundred extra quid, we will provide you unnamed benefits.*

Yuck.

Most people think writing notes is slower than typing anyway. All the while, we're getting more and more hooked on keyboards, and we may be so far beyond the point of returning to writing that the Tablet PC may not only occupy a graveyard slot but also a prominent position in the caravan of *quaint niche technologies.*

"We'll probably wait until we see an opportunity for this to become mainstream or where we can add value and jump in."

—Tony Bonadero, director of product marketing for Dell's Latitude notebook line, quoted on News.com, November 6, 2002

Lesson 1

Technologists' "simplicity" can raise actual TPPA because technologists can handle new technology much more "simply" than real humans can.

Lesson 2

The Tablet is a portion of a strategy—Microsoft has a wider view of how the Tablet fits into its overall plan.

Lesson 3

The Tablet is an expensive failure or application as opposed to a separate standalone technology.

Lesson 4

Announcements may actually indicate strategy instead of technology itself.

Lesson 5

Earthlings do not *want* to learn, nor do they enjoy learning.

FAILURE 5: WEBVAN

> *"Webvan has weathered numerous challenges, and in a different climate, I believe that our business model would prove successful. At the end of the day, however, the clock has run out on us."*
>
> —Robert Swan, former CEO, Webvan, in a July 9, 2001, company statement

> *"Here's a radical thought: The future of the online grocer market belongs to the grocery stores. They know the business, they can mix (sales) channels, and they can take their time."*
>
> —Whit Andrews, Gartner Group, quoted on News.com, July 9, 2001

> *"One unidentified venture capitalist described Webvan as a 'moonshot,' a daring, high-risk venture, and the reporter wondered alternatively whether the company would end up becoming 'the Internet era's equivalent of Waterworld,' a disaster so epic it becomes an American legend."*
>
> —Randall Stross, eBoys

Crisis: Low
TPPA: High

Background

Webvan promised to revolutionize the business of grocery shopping. During the late 1990s, Internet supermarkets were heralded as a means to attract explosive consumer demand. As recently as 2000, Internet supermarkets were still busy stocking shelves in one of the supposedly hottest segments of e-commerce. In fact, Webvan attracted more funding than any e-retailing company except Amazon.com, with high-profile, venture-capital backers such as Benchmark Capital and Sequoia Capital. The company would go on to raise $375 million in its IPO.

Amid the Internet bubble, Webvan came to be valued at $8 billion and touted an ambitious twenty-six-city plan. Bechtel signed a $1 billion contract with the company to build a string of high-tech warehouses for about $30 million each. But Webvan may have been ten or twenty years ahead of its time. And like many businesses born in the maniacal days of the dot-com boom, it aimed to get too big, too fast.

Webvan's problems were numerous: an overly aggressive expansion into new cities, rapidly disappearing cash reserves, trouble controlling expenses, difficulties managing the restructuring of operations, and an overly complex Web site. Although the company did manage to successfully cut operating losses, order volume didn't materialize and Webvan was unable to find the additional capital needed to keep operations running. Investors' early enthusiasm had pushed Webvan to expand more quickly than was wise.

As its financial problems intensified, Webvan found itself fighting another battle as well—with investors. The company fought to keep its shares listed on the Nasdaq stock market, going so far as to implement a 25-to-1 reverse stock split. Despite these efforts, the stock plummeted from a high of $30 during its 1999 IPO to a low of $0.06, and Webvan reported a net loss of $217 million and an accumulated deficit of $830 million in the first quarter of 2001. In light of rapidly disappearing cash reserves and a considerable drop in the volume of orders from customers, Webvan decided to shut down. In 2001, the company closed all operations and filed for Chapter 11 bankruptcy protection, just a year and a half after its remarkably successful IPO.

The Upshot

Webvan seemed staged to provide a service that would free up its customers' time—otherwise spent in a grocery store. Unfortunately, the company aimed to capture first-mover advantage, grow at lightning speed, and fork over gargantuan returns immediately. Conflict? It seems so. While Webvan's service had its merits, it didn't match to a nation-wide crisis. The crisis of shopping at an actual grocery store was simply too low in most markets and the uncertainty too great for changing important habits quickly. The perceived pain of adoption may have been larger than the reality, but Webvan wasn't financed in a fashion conducive to waiting out a slow and steady uptake by new customers.

Crisis

Webvan epitomized an era when venture capitalists made exceedingly large investments that were in complete contradiction to their previously stated investment philosophies. The idea of capturing the "first-mover" advantage contradicted the experience of the personal computer industry, in which the biggest share capturers were "late" entrants, Compaq and Dell. During the Internet mania, however, the passion to scale quickly was intoxicating. The fear of missing out was far larger than the perceived pain of losing money. In many cases the trauma of the "typical" venture capitalist experience was unleashed on the public equity markets long before previously tolerated.

> *"When Webvan began making incredibly aggressive investments, that's exactly what investors were telling it to do. Then Wall Street one day changed its mind, and Webvan suddenly found itself with an extraordinary amount of infrastructure and without the ability to get to profitability."*
>
> —Ken Cassar, *Jupiter Media Metrix*, quoted on Wired NEWS, July 10, 2001

Did venture capitalists have a crisis? You bet. Venture capitalists needed places to put the massive new amounts of money that were so easily raised. They felt a pressure to find *huge* winners. Some VCs were on the immortality bandwagon. The entire profession had for years been viewed as ethically equivalent to used car sales for the seeming preponderance of complex ownership details that got the better of the unsuspecting entrepreneurs that had sought funding. In the wake of the Internet mania, this same industry was seen as the home of the most brilliant folks on the face of the earth, so much so that books were rolling off the presses to capture and pass on some of that VC magic.

> *"One of the hallmarks of the dot-com crush has been the presumption that you needed to get big fast, which worked for Amazon.com and virtually no one else. The enormous infrastructure that Webvan thought to establish in multiple geographic areas just proved to be too great a cost."*
>
> —Whit Andrews, Gartner Group, quoted on Newscom, July 9, 2001

Another crisis was borne of the supposed first-mover advantage/ barrier. Everyone wanted it—and scaling quickly seemed to address everyone's needs. VCs needed to put ever more vast sums of money to work, and $5 million here and there would take too long. If only bigger financing rounds could be found so that $50 million at a clip could be deployed. Webvan was the answer to a VC's biggest problem during the Internet bubble—where to put all that money.

More so than not, the VCs were rolling the dice on public institution-sized situations that were occasionally in prestart-up stage. It made no logical sense unless we look at the Change Function from a VC perspective. There was a high crisis to put money to work.

That was great news for entrepreneurs. There had never been a time that *any* decent idea had *such a high chance* of getting funded.

It's unclear how many Internet start-ups were customer-centric entities or if VCs even encouraged such a focus. First-mover advantage was less about serving customers well and more about serving shareholders a gargantuan return. Few companies figured out that the latter was not possible without the former. It isn't surprising that many years later—as in, say, five to seven years later—eBay and Amazon are left but very few others.

So let's also look at Webvan from the end users' perspective. Did users have a crisis? Maybe. When my wife, Kelly, and I had triplets, we used a Web grocer for a bit—well, actually just once. What was it that folks like us—the parents of triplets—or even the rest of them were buying with Web-based grocery shopping?

Time.

That's a pretty good thing to be selling—Time.

James Gleick wrote an entire book, *Faster*, on the topic of the speed and rush building in our society—it makes a great bookend to Stewart Brand's *The Clock of the Long Now*. *Faster* was also a chronicle of cultural hyperactivity. Both books were among my quickest reads ever, perhaps demonstrating my own emergent and extensive hyperactivity.

According to Gleick, "Much of life has become a game show, our fingers perpetually poised above the buzzer," and he notes that people routinely press already lit elevator buttons in their frustration of losing a minute here or there. They also press eighty-eight seconds on the microwave instead of ninety—because it's a faster input. We're guilty of many such psychoses.

> *"I put instant coffee in the microwave and almost went back in time."*
>
> —Comedian Steven Wright

So Webvan was selling time, and lots of folks wanted time.

That's a good match.

We didn't know exactly how high grocery shopping ranked as a perceived time stealer, and Webvan was about to find out. The higher the rank of the alleged "time theft," the greater the likelihood that the crisis would outweigh the total perceived pain of adoption.

Total Perceived Pain of Adoption

So selling time is a great idea. Webvan had nothing to do with improving food quality or selection or price—just *time*. In fact, selection, quality, and price might suffer in the process. Would folks feel more imprisoned by a reduced choice in peanut butters, or were they more in crisis in wanting to save time—a classic weighing-machine dilemma.

The challenge for Webvan and others? To clearly enlighten the public about precisely what the service would buy you—time—and to somehow eliminate the many fears people had that gaining this "time" by changing buying habits would somehow come at a great cost.

One pain of adoption might involve resigning oneself to a lack of choice. At a certain point, experiencing or envisioning ten or twelve *I'll have to settle for Skippy instead of my beloved Peter Pan, but it saves me time so, yeah, I'll take the Skippy* trade-offs, one might throw in the towel.

But back to time itself. What would be the *actual* time saved? Webvan surely required some investment of time up front and each time you placed an order. And you'd still have to receive and put away the groceries. Which part of the process was *least* enjoyable? And . . .

- Would you have to be home to accept the groceries?
- What if you wanted to make changes to your order?
- Would they sell milk or would that be a trip to the store anyway?
- Would the produce be any good?
- What about meat and chicken and eggs?
- What about packaging disposal?
- Do they serve my area?

There were *a lot* of questions to ease folks through to see if online grocery shopping was for them.

Back to my own 1997 experience in online shopping. There was no local delivery back then, so we resorted to an online service that delivered from a warehouse in Dallas, Texas, if I remember correctly. Most of the items arrived safely, though we did experience the trauma of a broken jar of tomato sauce.

The main reason we did Internet–based grocery buying once and just once had to do with packaging. The supposed time saved vanished and then turned negative as we stood stuffing the Styrofoam protection—which clearly was not foolproof, as evidenced by the tomato sauce—into bag after bag after bag. Suddenly, *we* were responsible for the disposal of a nonbiodegradable mess.

But that's just us. In any case, from a total perceived pain of adoption perspective, groceries on the Internet suffered from a number of problems. Whether or not online delivery actually suffered these drawbacks is one issue—whether people *thought* it did was equally, if not more, important.

- Perception of unreliable delivery
- Perception of forced product selection—i.e., fewer choices
- Perception of unreliable order fulfillment
- Perception of a large set-up time
- Perception of a lack of cost savings
- Perception of an nonreturnables/brokens issue
- Perception of a need for a time-consuming live grocery trip anyway
- Perception of issues with package disposal
- Perception of list-making time eating into time savings

Webvan's demise had little to with the actual quality of its service, which generally received very favorable reviews. On ratings site Epinions.com, Webvan enjoyed an 89 percent approval rating, and ranked number one among all Internet grocery stores. Many parents of young children raved about how Webvan saved them from the hassle of regular trips to the grocery store.

It was the lack of customers that was Webvan's trouble—even as the company was building its empire of warehouses and fleets of delivery trucks, shoppers weren't signing up. By 2000, only 2 percent of Web users had bought groceries online in the previous year.

"On a given day, Webvan was delivering a grand total of 382 or-ders. The average order had 21 items, adding up to only $75."

—Randall Stross, *eBoys*

Webvan's Achilles' heel seems to have been that its first-mover, build-it-and-pray-they-will-come–business model was crafted by man-agement, venture capitalists, and other institutional investors. Also, Webvan did not remotely begin to respect the time required for people to even consider changing basic patterns of shopping when so many questions existed. It's hard to imagine a marketing program that would easily address the dozens of considerations folks might have in not jumping in.

Grocery shopping is such a vital part of life—as opposed to book shopping—that a change to that world without definitive benefit could have been perceived as risky. Once the Nasdaq collapsed, the peer pres-sure that had helped Amazon and eBay grow—and that might have guilted more people to get hip toward online grocery shopping—vanished.

A final consideration: The competition was changing. My own fam-ily history with groceries may have a few parallels with your own:

- My Sunday morning ritual of shopping at Stew Leonard's in Westchester with my kids is comical and enjoyable. I lead or am led by three children, and there's great entertainment value along the way.
- My wife and I often asked our au pair Ivan to do the shopping for us. How's that for efficient?
- Kelly has wrested back control of grocery shopping. I strongly suspect it is a result of her addiction to the experience of Whole Foods in White Plains. She actually seems to *enjoy* going.

Off-line grocers have been making great strides to offer higher qual-ity, more efficient, and more entertaining shopping experiences. Some-times they even throw in a lower price or two. Anyone who's visited Wegman's in Penfield, New York, will "get" how the experience has been altered. No technology play happens in a vacuum, and Webvan's competition was moving right along with it, in different and sometimes more compelling directions.

As a postscript, there are those that treasure their online grocery habit. I ran into some recently in Seoul—a highly concentrated, densely populated city. I recently found that Betty Wu Gallistel, of former UBS Global Tech Strategy fame, is also hooked in New York—another highly concentrated, densely populated city. Her choice—and that of many others—is Fresh Direct. In early 2004, Fresh Direct boasted twenty-five thousand transactions per week in New York. The "Fresh" in Fresh Direct proved to be actually fresh, and meant that the fear of receiving a brown, sad looking vegetable disappeared. But the real application Fresh Direct serves for Betty? Her husband, Adam, likes the pigs in a blanket that they sell. A "killer app," as they say. Hooked.

Over the next five to ten years we'll find out if demographic changes—a population growing more and more comfortable with digital lifestyles—will result in a different, bigger audience for online grocery shopping. Or we'll learn that the model only works in certain cities such as Seoul and New York—and that success will be limited to those who *didn't* think scale and first-mover advantage across fifty cities would turn a niche opportunity into the Next Big Thing.

Lesson 1

Investor crisis can greatly affect the business plan selected—watch out for private and public investors dictating the plan.

Lesson 2

Markets rarely adopt at microwave speed. The pace of adoption is nearly always far slower than initially imagined.

Lesson 3

"Perception" of pain can override any "absolutely" measured pain.

Lesson 4

Changing demographics may mean "failure" is merely time-specific. An idea can fail today and be a huge success tomorrow.

Lesson 5

Competitive solutions will develop while technology is evolving—alternatives are not static.

FIVE

FOUR MORE FAILURES

It is far too tempting to a student of technology to stop at a mere five examples of failure—there are just too many equally deserving candidates close at hand. I have spared you by picking on just five more for today.

FAILURE 6: THE ALPHA CHIP

> *"In short, the Alpha has a bright future. Although it currently has less than a 1 percent mainstream market share, I see it as the frontrunner in processors as the world begins to transition into full 64-bit environments. Although considered expensive by many, I am confident the 21264 will be a commercial success as market awareness spreads, speeds are increased, and prices are slashed."*
>
> —Slasher's Tech, July 14, 1998

Crisis: Low
TPPA: High

Background

In 1992, Digital Equipment Corporation introduced what it claimed was the world's fastest processor, code-named the Alpha chip. DEC's Alpha chip was a 64-bit microprocessor that ran at 200 megahertz, far faster than any other microprocessor on the market. By comparison, the Intel Pentium chip launched the *following* spring ran at a mere 66 megahertz. Alpha was born out of an earlier failed project that had been designed with the intent of releasing a new operating system.

Although the failed project, named PRISM (parallel reduced instruction sct machine), was ended in 1988, it was clear at the time that subsequent generations of new chips would offer much better price/performance ratios. Consequently, the decision was made to upgrade the design to a full 64-bit implementation from PRISM's 32-bit. This 64-bit architecture became the Alpha project.

Around 500,000 Alpha-based systems were sold by the end of 2000, but for a long time the Alpha chip was only used for 64-bit graphics programs that required brute processor speed.

There were also several barriers to the widespread adoption of the Alpha chip. First, the 64-bit proccssor couldn't natively and effectively and seamlessly run most of the current 32-bit or 16-bit applications. The Alpha chip also required potential users to learn a new operating system because the highly popular Windows operating system didn't work with a 64-bit processor. Finally, Alpha computers were also extremely expensive, costing around $6,000. Compaq eventually purchased DEC, and ultimately decided to completely phase out computers using the Alpha chip by 2004 in favor of Intel's Itanium chip.

The Upshot

DEC's 64-bit Alpha chip was technologically brilliant. But brilliance and innovation alone do not create new markets, despite how much technologists would like to believe that they do. DEC failed to create a network—or an environment—in which its Alpha chip could be embraced. The technology ecosystem failed to support Alpha, so while the crisis was low, the total perceived pain of adoption—the lack of operating systems and applications to support Alpha—was quite high. The biggest fear for users was that they would embrace Alpha and then wind up stranded on an island.

Lesson 1

A lack of crisis and a high TPPA are hard to overcome. With the Alpha chip, there was a large audience that wanted more speed in processing but did not want any of the attached struggles.

Lesson 2

The Alpha depended on full ecosystem support—a dangerous proposition. Swapping out just a portion of a food chain is one thing but demanding changes through the whole ecosystem is a tall order.

Lesson 3

Even early adopters sometimes need others to follow quickly—the benefit of widespread adoption is a reduction in many aspects of the total perceived pain of adoption. In Alpha's case, most folks wanted certainty that Alpha would be a de facto industry standard but didn't want to be the first kid on the block to adopt.

Lesson 4

Users weren't convinced DEC would remain committed—if Alpha didn't take off, DEC would be foolish to continue to support it—no one wanted to be stranded on an island one day.

Lesson 5

Who wants to migrate applications unnecessarily? What a pain in the neck! This falls into the category of "if it ain't broke don't fix it."

FAILURE 7: ISDN

"But ISDN Is Still Difficult to Network properly, it still doesn't work well, standardization efforts are just getting started, and there are too few Internet Service providers around that offer ISDN access. If you pass all these hurdles, it can be a very effective way to access the Internet."

—David Strom, "Is Still Difficult to Network," December 31, 1996, www.strom.com

Crisis: Moderate at End User
TPPA: High

Background

First offered in 1980, ISDN—Integrated Services Digital Network—is a digital telephone line that was designed to create a "next generation" telephone system that would integrate voice and data into one connection. ISDN phone lines consist of anywhere between two and thirty separate wires or "channels" depending on the type of service, with each channel simultaneously offering a connection of 64kbps (kilobits per second) that can transfer fax, voice, or data at approximately five times the speed of a standard 28.8kbps connection. These 64kbps channels can be used independently or combined to give a larger bandwidth, meaning that ISDN is well suited for handling many concurrent connections or calls conducted from a single location.

ISDN took more than a decade to become widespread and did eventually have a slight up-tick in popularity as pricing reached the forty to sixty dollars per month range and connection hardware expenses fell, but ISDN service has still been almost completely displaced by more common broadband Internet services such as DSL and cable modems that are faster, less expensive, and easier to set up and maintain.

The Upshot

ISDN's failure resulted from the high total perceived pain of adoption that confronted users considering the service. Yes, it was a nightmare for users to have ISDN installed, but telecom service providers—in not properly marketing the service or providing anything resembling customer service—ultimately killed the technology before much of its promise was realized. For monopolists, there is simply nothing wrong with the status quo. So when it comes to depending on a slow-moving behemoth to bring pioneering technologies to the market, don't hold your breath.

Lesson 1

Crises can be instigated through awareness campaigns—in the case of ISDN the telcos did next to nothing to raise awareness of the service. Great marketing helps folks have ah-hah moments when they see why they might want something that they previously didn't even know or understand. The telcos have been the antithesis of great marketers.

Lesson 2

The conduits to change *can* halt technology change—since ISDN technology needed to flow through the telco channel there were inordinate obstacles to overcome, given the telcos' history of *not* deploying next-generation technology rapidly.

Lesson 3

Potential crisis inducement can be overwhelmed by high TPPA—since the marketing was horrendous, the high total perceived pain of adoption dominated the possible adoption of ISDN.

Lesson 4

Monopolist crisis is rarely about user experience—monopolists become very far removed from empathizing with the end user.

Lesson 5

Monopolist crisis relates to a threat to the monopoly itself—it's about the telcos—not the users—when the competitive environment is thin. When competition heats up, it is very difficult for an organization to shift its culture to once again think about users and compete.

FAILURE 8: ASPS

> *"Companies that subscribe to an ASP's services use the applications just as consumers use telephone voice-mail that's offered through, say, AT&T or Verizon. The voice-mail technology does not exist in the person's house; it lives at the phone company. The user simply pays the phone company a monthly bill to access its technology. This saves the consumer the cost of buying, maintaining, and replacing an answering machine."*

> —CIO.com

Crisis: None
TPPA: Very High

Background

Application Service Providers (ASPs) were marketed as a completely new way to sell and distribute software and services. The ultimate dream of ASPs was to allow users access to any of their applications and data whenever and wherever. The phrase *software as services* is a close cousin to ASPs—with the mild distinction being that ASPs would be newly formed entities while established packaged applications vendors were expected to offer similar software as services across time in hopes of creating a more consistent business model.

Simply put, ASPs were companies that provided software applications and services over the Internet. These ASPs owned, operated, and maintained the software applications and the servers that ran the applications. Pseudo-examples of ASPs were Web hosting and e-mail facilitation. The main appeal of ASPs was supposed to be that software and software upgrades had become increasingly complex and expensive—subscribe to an ASP, the argument went, and those costs went out the window.

The ASP model sounded extremely appealing—especially to small businesses and start-ups—because it offered the potential to dramatically reduce costs, an extremely short set-up time, the ability to greatly reduce expensive IT head count, and reduced maintenance costs.

ASPs would also—some would say mainly—positively impact the typical software business model. Typical packaged applications companies suffered from a licensing fee structure that made sales lumpy and caused stock prices to be volatile. Investment bankers convinced the industry that investors would pay more for the steady flow of revenues the ASP model represented. But that was easier said than done. While early projections of worldwide spending on ASPs were in the $25 billion range, actual spending has never reached even a fraction of that amount. As we roll the tape the entire way to 2006, the revenue estimate for the only widely known purveyor of software for lease—Salesforce.com—is just $450 million. That's great for Salesforce.com, but a drop in the bucket compared with expectations.

The Upshot

When ASPs first hit the scene in 1999, was there a user crisis for enterprises to access their core software applications over the Internet from a third party provider on a rental basis—to have these applications sitting somewhere else other than their own offices? Nope. There still isn't much of a crisis, although Marc Benioff's Salesforce.com has been able to convince users to give it a try.

Did the ASP model offer a benefit for software vendors? Sure. It smoothed out lumpy, back-end-loaded enterprise software sales into smoother subscription-like revenues. But why should enterprises change their behavior patterns to facilitate improved profitability and financial transparency for vendors?

Lesson 1

"Software as services" was on everyone's agenda except real users'—to say there was no crisis at the enterprise level would overstate it a touch as it might suggest that those users even gave it a moment of thought.

Lesson 2

Solutions that "fix what isn't broken" usually fail—end users have many other things to think and worry about.

Lesson 3

In 1998, the crisis that ASPs solved didn't exist for users.

Lesson 4

In 2005, many end users' crises have opened the ASP door a tad. The culture has changed at the enterprise level. Today the CIO is far more open minded to the idea of outsourcing some applications. The interest is still minor, however, as it typically still takes a backseat to a desire to control as much as possible.

Lesson 5

It's crucial to "get" what is on end users' agendas and cut through the hype emanating from vendors' agendas.

FAILURE 9: ELECTRONIC EXCHANGES

> *"Is it absurd to believe that in 2001 the electronic transactions of Brazilian companies will total an amount greater than Uruguay's GDP? Would you believe that by 2005, total e-transactions in Brazil will be almost two-thirds of the entire Colombian economy? With the help of the Internet, business-to-business e-commerce can indeed achieve this performance, growing from $15.3 billion in 2001 to $51.7 billion by the end of 2005, a compound annual growth rate (CAGR) of 35.6 percent."*
>
> —The Yankee Group, "The Brazilian B to B Economy Emerges,"
> December 1, 2001

Crisis: Temporarily High at CEO
TPPA: Temporarily Low at CEO

Background

Electronic exchanges are virtual business-to-business—B2B—electronic trading networks on which companies conduct business over the Internet in a many-to-many (instead of one-to-one) fashion. Most of the first generation B2B exchanges were open public marketplaces with business models based on the collection of a small percentage fee on all transactions conducted. Proponents of electronic exchanges argued that companies would benefit from economies of scale on the purchase and sale of materials and from exchanging specs, designs, and blueprints with suppliers with whom they would otherwise not interact.

Electronic exchanges initially captured considerable interest, especially among smaller manufacturers, because they could be accessed with a normal Web browser instead of expensive, proprietary, one-to-one electronic data interchange, or EDI, software. As a result, electronic exchanges were created for a number of industries, including the auto-

motive industry, the PC industry, the chemical industry, and the petro-
leum industry.

The total worldwide value of goods and services purchased by businesses through e-commerce solutions was projected by market research
firm IDC to reach $5.8 trillion by 2006, and by 2000 there were over one
thousand electronic exchanges in operation. But then, on the verge of
success, interest collapsed.

The major problems facing electronic exchanges included integrating different e-procurement platforms effectively, including those provided by Ariba, Commerce One, i2, Netscape, Oracle, and SAP AG.
Although electronic exchanges attempted to improve their interoperability, the initial technical challenges were followed by the additional
difficulty of attracting paying customers and suppliers—many of the
biggest were already hooked on EDI. When the pressure of the Internet
mania subsided, the pressure to deploy subsided as well.

The Upshot

The technology community fell hard for the concept of electronic business-
to-business (EB2B) exchanges and the companies that offered the technology to build them in early 2000. The sense of crisis at that time was
so palpable, it reached near fever pitch. CEOs of global corporations
rushed out press releases about streamlining their global supply chains
by working with the Aribas and Commerce Ones of the world. But we
mistook crisis to announce for crisis to deploy. Yes, there was pressure
for CEOs to seem hip to the New Economy and have an Internet strategy. But there was little actual pressure to turn this vision into a reality.
The high TPPA of actually building an EB2B exchange—as clever as it
was—was never actually put to the test because as the markets turned,
EB2B went from golden child status to disavowed stepchild in a flash.

Lesson 1

Announcements to deploy a new technology do not equal actual
deployment—a press release does not equate to an actual event. In the

case of electronic exchanges, the PR announcements quieted management critics who feared their companies had no Internet strategy. The announcement was the easy part and provided the most relief—follow-up activity was challenging.

Lesson 2

Altering incumbent EDI systems carried such a high TPPA that many folks wondered what the true benefit was of moving away from a system that already worked.

SIX

A FUTURE WINNER: FLAT PANEL DISPLAY TV

Crisis
Today: Moderate
In 12 Months: High

TPPA
Today: Significant
In 12 Months: Moderate

Flat panel display technology promises to provide sharper pictures when used in conjunction with high definition television technology or high definition DVD content. It also promises to save space relative to current technology. But what flat panel technology promises and delivers more than anything is coolness.

Background

For the past seventy-five years, the vast majority of televisions have been built using cathode ray tube (CRT) technology. In a CRT television, a gun fires a beam of electrons inside a large glass tube. The electrons excite phosphor atoms along the wide end of the tube, which causes those atoms to light up and produce the television image. Cathode ray tubes produce crisp, vibrant images, but they are bulky, heavy, and in order to increase the screen width in a CRT set, the length of the tube must also be increased. Flat panel displays are an alternative to CRT TV. There are currently three main contenders in the flat panel display space: plasma, Liquid Crystal Display (LCD), and Digital Light Processing (DLP).

Plasma displays have wide screens, but are only about six inches thick. The basic idea of a plasma display is to illuminate tiny colored fluorescent lights to form an image. Each pixel is made up of three fluorescent lights—red, green, and blue. Just like an old-fashioned cathode ray tube (CRT) television, the plasma display varies the intensities of the different lights to produce a full range of colors. The main attraction of plasma technology is that you can produce a very wide screen using extremely thin materials. Because each pixel is lit individually, the image is very bright and looks good from almost every angle.

Many everyday items contain the second form of flat panel displays—LCDs—including laptop computers, digital clocks, watches, microwave ovens, and CD players. The technology involves the passage of light through liquid crystals that generate differing results when an electric charge is either applied or not applied. While liquid crystals were discovered in 1888, eighty years passed before RCA made the first experimental LCD in 1968. Since then, LCD manufacturers have steadily developed variations and improvements on the technology. LCDs are thinner, lighter, and consume less power than CRTs. The main drawback is that the display size is limited by quality-control problems. To increase display size, manufacturers must add more pixels and transistors. But as they increase the number of pixels and transistors, they also increase the chance of including a bad transistor in a display.

The third and final flat panel technology, DLP, is a projection technology predicated on the use of so-called micromirrors. In contrast to conventional TVs, projection TVs form a small image on a device inside the projector and then shine that image onto a large screen. DLP TVs use a Digital Micromirror Device (DMD) chip that has anywhere from eight hundred to more than a million tiny mirrors on it. When voltage is applied to the DMD chips, each mirror can tilt +10 or −10 degrees, representing "on" or "off" in a digital signal. Light hitting the "on" mirror will reflect through the projection lens to the screen. Light hitting the "off" mirror will reflect to a light absorber. By using a color filter wheel between the light and the DMD, and by varying the amount of time each individual DMD mirror pixel is on, a full-color, digital picture is projected onto the screen. Projection TV technology can create large screen sizes at a reasonable price.

The Market

According to industry research firm Display Search, in 2004, flat panel display televisions accounted for just 9 percent of global television sales. In Japan, flat panels already account for nearly 30 percent of sales. The market should experience a critical inflection point during the next two years, perhaps even beyond Display Search's healthy expectation that over 25 percent of 2006 sales in the United States will be flat panels. Before long, it's likely that the words *flat panel* will be unnecessary, as *television* will mean a *flat panel television*. Peer pressure will have a profound impact on the marketplace and obviate the need for prices of panels to drop to the extent that supply and demand modeling might suggest.

The Upshot

That flat panel TVs may be considered "boring" by investors is a direct result of how often this technology has been mentioned in the press. Nevertheless, the technology may be at the most critical point of the so-called S-curve of adoption where sales really take off, in a scale rivaling the major product trends that dominated the 1990s. This is an upgrade cycle of the one billion-plus global television market, not to mention that flat panels also go with desktops and laptops as well as the emerging public signage market. The total perceived pain of adoption on the basic flat panel television sale boils down to price, as very little education is required. It rarely gets so good in the technology world! But a low total perceived pain of adoption itself doesn't generate purchases. There is a heavy word of mouth and word of sight "coolness" experience factor that may drive sales of flat panel TVs even more effectively than billions of dollars spent on advertising. Flat panels are cool. Buying them implies immediate returns on coolness. Peer pressure will build fairly rapidly during the next couple of years and people will spend more money than they ever dreamed of on televisions and forget that electronics stores ever sold anything else.

Crisis

> *"The new technology—HDTV, for high definition television— was recently described by a Federal Communications Commission advisory panel as 'an economic opportunity of almost unparalleled proportions.'"*
>
> —Norm Alster, "TV's High-Stakes, High-Tech Battle,"
> *Fortune* magazine, October 24, 1988
> (Yes, that quote was from 1988. No typo there.)

Your first reaction to "flat panel display" may be *Duh! Everyone knows this!* And to a large extent, that is true. Technologies rarely set up much better than this one.

There are at least four active markets for flat panel display technologies. The focus of this argument is just one of them: flat-panel display televisions. The others are the still fast-growing laptop market, the desktop transition opportunity, and the burgeoning but still small public signage and display space.

It's early yet: In 2004, only 9 percent of televisions sold globally were flat panel displays. Japan? 28 percent. The United States? Approximately 15 percent. In western Europe, mind you, the figure was well under 9 percent.

So, what's the level of crisis for flat panel televisions? Today, here in late 2005, the "general" crisis level is still modest, but we care only so much about the general crisis level in the early stages of technology adoption. We care, instead, about identifying enough early movers to get the ball rolling.

Who are the folks that initially have a higher crisis level than others? The profile varies modestly around the globe. In the more penetrated

Japanese market, *space saving* is a big factor. Apartment sizes in Tokyo are much smaller than even in London, Paris, or New York—and although pricing on flat panel TVs tends to be higher in Japan, a flat panel television is still particularly attractive. In a tour of the massive Akihabara electronics district in Tokyo way back in November 2003, you'd be hard-pressed to find a lumbering CRT television set amid the dominance of flat panels. You'd have trouble finding large-scale television sets at all—CRT or flat panel. The sweet spot for sales folks was the twenty-one-inch screen that was workable in the Tokyo apartments. In a global context, "space savings" is a bit of a start but still a fairly limited "crisis."

The bigger selling point? Ask a salesperson at Best Buy or Frye's or in Akihabara why people buy flat panels. The answer will invariably be the same.

They're cool!

Intel CEO Paul Otellini seems to think they're cool. He had twenty-nine of them flashing three distinct messages on stage at the 2004 Consumer Electronics Show in Las Vegas. Flat panels were already the established fashion in Tokyo in November 2003. The holiday windows in Saks Fifth Avenue in Manhattan now feature flat panels.

Tech culturalist Stewart Brand developed a way of thinking about the pace of change across a variety of situations. He offers seven stages of a culture relating to the speed of change. The slowest change occurs in nature. The most frequent change occurs in the uppermost band in his construct—the band that says "fashion/art."

Flat panels may indeed be a fashion. When we buy cars, clothes, music, deodorants, beer, or cereal, advertisers prey on the association consumers have between the things they buy and who they are. If you buy Heineken, you're "sophisticated." And so on.

In that vein, if you buy a flat panel, the *implicit promise* is that you will be cool. You may or may not be blown away by the picture since Earthlings are exposed to crisper and crisper pictures in a variety of locations every day, and so, while in 1988, it might have been a revolutionary picture, today it really ain't. Nonetheless, for $2,000 someone might want to say they are indeed "blown away," since spending several thousand dollars just to be cool doesn't seem anything to brag about in and of itself.

> *Choice A: "Well, I am really insecure, so I bought a flat panel to feel like I fit in and because maybe people will like me more when they come over to watch the game."*

> *Choice B: "The picture is amaaaayyyyyzzzzing."*

> *Choice C: Combination of A and B*

If you're feeling like this whole digital revolution is passing you by because you're an analogist and scared of researching your vacation online or setting up an iPod—though you *really* do like the idea of having your five hundred CDs "with you" at all times—you can in an instant get on board and feel a whole lot better about your digital self with one single purchase. As the Barenaked Ladies sing in "Shopping," "Everything will always be all right . . . when we go shopping." on their CD *Everything to Everyone*

So you buy a flat panel display. Voilà—you're part of the digital age—presto chango. Will the "cool" factor fade anytime soon? I doubt it. While in time, I do suspect it will ultimately fade from "cool" to merely "standard issue" as in *you just have a flat panel TV, because that's what a TV is!* "Cool" will find a new place to live. But that's not what I'd call a problem for the makers of flat panels. Coolness doesn't really begin to fade until a high penetration rate, once everyone has the thing.

Flat panel televisions have been considered very cool for quite a while now. The first flat panel display was developed by Sharp in 1988. The past two CES conferences in Vegas and the past several Computex shows in Taipei have been bathed in the glow of flat panel emissions at every turn.

The "coolness" of a DVD player has fallen dramatically. DVD players are now a given in society—nothing to get excited about. Ditto for most

PCs. You don't often hear, *Hey, I got a Personal Computer! (subtext: I am so cool!)* unless, perhaps, someone has switched to a Mac or bought a Mac-Mini. Were people bragging about PC purchases ten years ago? Yep.

Takeaway Point 1:
Our relationship with technology—and its coolness and what owning a particular piece of technology says about who we are—is vital in adoption cycles.

The real pressure point in adoption is a result of . . . *peer pressure . . .*

PEER PRESSURE

Peer pressure comprises a set of group dynamics whereby a group in which one feels comfortable may override personal habits, individual moral inhibitions, or idiosyncratic desires to impose a group norm of attitudes and/or behaviors.

—wikipedia.com

Why does peer pressure kick in? People don't want to be isolated. People fear that if they don't fit societal standards they will be left out. Subconsciously, they would rather follow a herd than risk abandonment.

THE IDIOCY FACTOR

What Malcolm Gladwell superbly and neatly popularized as *tipping points,* or the inflection points of the so-called S-curve of adoption, we have named "the point of idiocy." This is the point when a growing number of people feel like idiots if they have to fess up to not owning——(*fill-in-the-blank*) technology.

To be clear: people aren't *actually* idiots—they just *feel* like idiots.

"No, I still don't have a cell phone."
"No, I still am using a VCR."
"No, I still have a paper-based calendar."
"No, I am still playing 8-track tapes."

The point of idiocy is a "when." It has nothing to do with the rationality of doing this or that—it has everything to do with "when" a certain portion of a community or subcommunity has adopted a view and taking the opposite view causes pain of some sort.

The point of idiocy is when one's day-to-day interactions with humans become a stronger influence on oneself than any print, screen, or radio advertising. My emphasis on the point of idiocy is not to suggest that consumers don't also find the merit of using one new technology or another, but rather that the peer pressure helps put the issue at the top of the agenda when it might not have otherwise gained much attention. In the 1990s, the point of idiocy showed up at cocktail parties when folks "needed" to pull out a PDA or mention their day-trading habits in order to fit in.

Peer pressure is about belonging to something larger than oneself. But there's an admission fee, of sorts, to gain entry to that part of society. Following fashion and fad can be part of that price.

So . . . do people buy technology just because it's better?
Nah!

Do technologists think if they build something that is just flat-out better they will create a market?
Yes.

Are these two perspectives at odds?
Often.

Who decides who is right?
The customer is always right. The customer has the money that the vendor is interested in securing in order to establish a business. But the customer might well be indifferent to what the vendor offers.

How close are we to the point of idiocy for flat panel televisions? Penetration of flat panel TVs in the 2004 sales mix was 9 percent. The point of idiocy normally starts to show itself in the 10 to 15 percent range, and it becomes overwhelming just above that. In 1999, seven million DVD players were sold in the United States—that was approximately 6 percent penetration. (Seven million of a total of 110 million households.) In 2000, thirteen million units sold or 12 percent penetration. Then, within two years, forty-six million households owned them—42 percent penetration.

12 percent to 46 percent inside two years.
Is that the magic formula?
It would be cool if we had a magic little formula: 12 percent to 46 percent.
Too bad there's no magic formula.
. . . but the area around 12 percent makes sense for the peer pressure to start growing . . . businesses that appropriately play pre-5 percent and pre-12 percent and then go to an altered strategy from 12 percent on will be well served by doing so . . .

It seems likely that we'll be closing in on similar penetration for flat panel televisions during the next two years, and there's a good chance that the holiday seasons of 2005 and 2006 will cement the shift. Vendors will probably find themselves surprisingly short of supply.

SIGNAGE

There are other factors supporting flat panel adoption. The first is a separate form of free subliminal advertising above and beyond peer pressure. The adoption of flat panel screens by retailers around the planet establishes "cool" in a similar way to what other technologies achieve with celebrity placement. Flat panels are attention-getting media, and retailers know it. When The Gap uses a flat panel, they're advertising not only their clothes but also the concept of flat panel television itself.

You know how we say that kids are attracted to images, and so we place them in front of a monitor as a quasi-babysitter at times? Well, kids grow up into adults—horrors—like us, and then we all stare at the goofy news blurbs on the flat panel riding the elevator up the 555 California building in San Francisco. We stare at the massive flat panels in Seoul. We stare while waiting to order Big Macs at McDonalds in Hawthorne, New York. Subway entrances are decorated with them, as are the windows at Saks. New York City health clubs have them in their windows. We get a subliminal message that the flat panels are cool while we ride a stationary bike at the Metropolitan Athletic Club or buy bananas at Wal-Mart. Wal-Mart's in-store advertising "network" is apparently the fifth largest television network (in terms of viewers) in the United States.

But the craze has really gone over the top when I have to ask my dental hygienist, Joannie, to turn off the flat panel during my teeth cleaning because, after all, the only reason I go to the dentist is for the peace and quiet I otherwise only find on airplanes. When my sensational dentist and technology investor Thomas Magnani moved spaces, maybe he decided his customers would have lower total perceived pain of adoption during gum scraping if they could watch Beyoncé on MTV. He does use the flat panels for medical reasons as well. Big pictures of teeth and gums . . .

> "Here in the Houston suburbs, Banana-Vision has arrived. That's the industry nickname for the 42-inch high definition LCD monitor installed directly over a pyramid of bright yellow bananas in the produce section of the local Wal-Mart store. . . . According to Wal-Mart and to an agency that handles its ad sales, the TV operation captures some 130 million viewers every four weeks, making it the fifth-largest television network in the United States after NBC, CBS, ABC, and Fox."
>
> —The New York Times, February 21, 2005

HDTV AND HD-DVD

Discussion of HDTV really belongs on the total perceived pain of adoption side of the flat panel ledger, but buyers seem to be just willing enough at the point of a flat panel purchase to consider that *at some point* it will all be *figured out* and buying an HDTV-enabled set gets you that flat panel you long for today plus an upgrade path into the future. Salespeople can use "digital" and "HDTV" in the pitch and then deflect an awkward moment when the buyer asks "will this get me HDTV?" by saying, "It's HDTV-enabled, so when HDTV is available in your area this flat panel will work."

This reduces the buyer's fear that they could be making a mistake, buying the wrong flat panel for $1,500 with the chance they'll be back buying a new one a year down the road. Score one for the salesperson!

> *"Things start to get more complicated when rationalizing the setup and connectivity conditions to view, record, distribute, and connect digital TV signals, in other words 'a HD system,' not just a HDTV. Complexity and confusion have the potential to affect timeliness of the transition to HDTV because confused people are usually reluctant to buy a relatively expensive product."*
>
> —Rodolfo La Maestra, *HDTV Etc.* magazine,
> February/March 2005

The upside to the HDTV talk is that an upgrade cycle to accommodate digital transmission will further inspire households—that normally upgrade televisions every eight to ten years—to consider whether the time is suddenly right in order to become a compatible household. "HDTV" may vie for the title of the *simplest* and simultaneously *most complicated* technology to adopt.

The simplicity is really easy to grasp: A much better viewing experience. The complexity? When I asked David at the Samsung Experience store, "What else do I need to do after I buy one of your flat panel television products in order to get the best possible HDTV picture?" he told me I had to figure out my "content feed." And that's where it gets complex. Instead of just buying a television, I now need to:

—Buy an HDTV-ready television.
—Get a set-top converter
—Establish a relationship with a provider of HDTV content.

Salespeople at consumer electronics stores selling flat panel displays have no real idea about how the whole "content feed" thingamajig really works.

Flat panels can also work with newly enhanced DVDs that should be on market before too long. Not surprisingly, two competing standards—HD-DVD and Blu-ray—are still battling it out. Might these high definition DVDs encourage flat panel purchases while we wait for our friend David at the Samsung Experience to simplify the HDTV buying experience to something less than its current fiasco? Sure.

> *"HD-DVD is just an extremely elegant extension of today's existing technology and it arguably can sell a lot of (discs) over the next three or four years. But looking to 2009 and beyond, Blu-ray is really the technology that gets you the higher storage that you need."*
>
> —Gerry Kaufhold, research firm In-Stat

Total Perceived Pain of Adoption

The short answer? The TPPA is *very* low.

The absolute price of a technology tends to take up only about 10 percent of any total perceived pain of adoption calculation, the remainder being represented by considerations like learning something new, failing to learn something new and feeling like an idiot along the way, waiting on a help line on a Saturday afternoon and so on. It would be nice if price represented a much larger portion—because with fewer of the more complicated factors, the forces of Moore's Law would be much more effective in lowering the overall pain of adoption. If price is the only issue to contend with, that's great news—it only boils down to a function of time at that point.

Unfortunately, there are many elements that need to adjust with the introduction of change, and it's almost never just about the price. Not surprisingly, technologies that affect fewer elements of an ecosystem stand a greater chance of more rapid adoption.

Doug Engelbart invented the mouse. He's been on a crusade to make technology usable for over fifty years. Here's the way he put it to me: "The technology is always way ahead of the ability of the system to absorb it."

Or, if you'd like a second opinion: "*We don't need your education.*" That's from Roger Waters and Pink Floyd on their album The Wall.

But this is a highly unusual circumstance. Flat panel televisions require virtually no education, and there seems to be no need to calm any fears that some form of education is required. Most everyone feels that if they go to Akihabara or Best Buy and plunk their money down, they will receive an object that they will take out of a big box and plug into a wall socket, and—voilà—they will have a really cool television. With the exception of some remote controls, folks aren't afraid of televisions. The salesperson's job is to ensure that they don't mess up the deal by bungling the inevitable questions about HDTV.

It's just a television. It's not an iPod. It's not a Personal Video Recorder. It's not an entertainment PC. There is no need for a manual. It is simply, reliably, easily, dependably a television.

Sharp inflection ahead!

The slope of the inflection will likely be extremely sharp. Given the immense social pressures likely to mount during the next two years, the price of flat panels won't need to drop to CRT-comparable levels for a New England Patriots fan to scurry out and become digital before the next Super Bowl they might win.

When wistfully reminiscing about the nineties and the fabulous eight or nine major inflection points—points of idiocy—we experienced: database to ERP to PCs to LANs to the Internet to CRM to Wireless to

Y2K and Telecom Deregulation, we lament that there is really only one comparably massive technology likely to overwhelm the tech industry in the next five year.

That technology is flat panel. The stars couldn't line up more perfectly. Vast end markets are already established—from PCs to notebooks to televisions to public signage. There is, conservatively, an installed base of easily over 500 million PCs and one billion televisions. Flat panels are a cool technology with the potential to create large social pressure. They require near-zero education. The technology is viewed as simple, easy, reliable, and dependable. And there are a couple of content drivers— HDTV and High Definition DVD—to help accelerate an already exciting trend. And Moore's Law is already at work. I don't expect to see such a powerful set of forces anywhere else in technology anytime soon.

SEVEN

THREE MORE FUTURE WINNERS

I can apply change thinking to nearly everything under the sun and am tempted to blab on and on and on about this tech winner or that tech winner, but I only subject my clients and regular Waypoints followers to such cruelty. Here I will limit myself to just three more future winners.

FUTURE WINNER 2: MOBILE ENTERPRISE E-MAIL

> *"Our vision for mobile e-mail extends well beyond the top executives and traveling salespeople who use their BlackBerry or Communicator type of devices at airports. We believe that mobile e-mail will penetrate across the entire workforce so that everyone will benefit from it in just the same way as the mobile phone has unleashed great productivity gains."*
>
> —Pekka Isosomppi, Nokia-Enterprise Solutions

Crisis: High
TPPA: Low

Mobile e-mail in the enterprise promises to assist management in creating an always-on environment in which an entity has access to all resources—including people and their knowledge and problem-solving ability—at all times. The technology promises this while empowering the workforce to be ever more mobile. Employees are promised an always-on connection to anyone in the network at any time as well as one of the ultimate personal networking devices.

Background

Mobile e-mail is a logical extension of conventional e-mail, which has become critical in the running of any sort of enterprise over the last few years. E-mail in the office has changed the way businesses operate by eclipsing or complementing traditional communication methods like the fax, telephone, and paper forms. But flexible, mobile organizations need more than e-mail facilities in the office, and mobile e-mail is the next stage in boosting efficiency. Mobile e-mail will be a key driver of adoption of wireless technologies by corporate enterprises, since companies typically start their mobilization strategies with mobile e-mail and then layer on more complex applications using the same devices. The main selling points of mobile e-mail include increased worker productivity, flexible working practices, and relatively simple deployment.

Despite the current buzz around mobile e-mail, the market is still relatively untapped. The Yankee Group estimates that there are currently 50 million mobile workers in the United States alone, but the market leader in mobile e-mail, Research in Motion, had a mere 3.5 million United States subscribers in late 2005. Mike Lazaridis and Douglas Fregin founded RIM in 1984 as the first wireless data technology developer in North America. Ten years later, RIM introduced the first BlackBerry hardware device, which was followed a year later with the introduction of the BlackBerry Wireless E-mail Solution and Enterprise Server Software.

The BlackBerry is widely regarded as a pioneer in mobile e-mail with its simple, always-on features, and the recent licensing of BlackBerry software to other hardware providers' devices enables companies to have mobile devices with far greater potential to support other applications. Nokia has also entered the mobile e-mail domain, recognizing the likely importance of mobile e-mail put into more and more devices.

Nokia recently released its own e-mail messaging solution, Nokia

One, a competitive solution to RIM's Black-Berry. Microsoft is also focused on the mobile e-mail space. Although the entrance of these competitors into the market indicates the potential growth for mobile e-mail, the largest obstacle facing most mobile e-mail providers is the relatively costly devices and service. Perhaps the most significant element in supplying corporate mobile e-mail solutions lies in bringing down the total perceived pain of adoption for all parties involved: service provider, CIO, CFO, and the actual end user.

BlackBerry addresses a market where every member of the ecosystem has a crisis. CIOs and CFOs are on the hook to make their organizations more mobile, more virtual, and more and more "always-on." Employees only reluctantly want these attributes as well. Service providers are desperate for new revenue.

The brilliance of BlackBerry is that it reduces pain of adoption for all parties. Blackberry has nearly all the mobile e-mail subscribers, making it a brain-dead decision for CFOs and CIOs. BlackBerry has proven itself with all the reference accounts as solid on security and support. Why bother taking a chance on an upstart? If it falls short, you will be viewed as an idiot and will be fired. Users take what management says and are quite happy toting a BlackBerry as a sign of self-importance while zipping around airports. Service providers? Over one hundred service providers offer BlackBerry because they want to sell a solution and their customers have a low TPPA themselves. BlackBerry is the de facto standard.

Mobile e-mail solutions will likely gain more functionality than they already have in terms of intelligent viewing, content rendering, and filtering that give mobile users the gist of the attachments and documents they need to work on while on the move. And encryption and security measures are becoming available to better meet CIOs' requirements.

Market

As of late 2005, there were 4.3 million e-mail users subscribing to Research In Motion's BlackBerry service, while there are approximately 150 million corporate e-mail users worldwide. While the equity market may have concluded that the market is already saturated as financial types pass each other on the streets of New York City and eye each other's devices at analyst meetings around the world, we're probably still eighteen to twenty-four months away from hitting a major inflection point on a much larger scale.

The Upshot

Some readers may assume that everyone already has a BlackBerry and that the inflection point—the pandemonium—could already be over. But with only 4.3 million BlackBerry subscribers in late 2005 and 50 million "mobile" workers in the United States alone, saturation hardly seems to be the case. People love communicating and being able to connect wherever they are. Enterprises love knowing that their employees can be more accessible than ever. But the icing on the cake is that the total perceived pain of adopting a technology that mimics an application most people already know and are familiar with—e-mail—is very low. Mobile e-mail may be one of the biggest winners as a result.

Crisis

The crisis level for mobile e-mail in the enterprise is quite high. Users want to be connected to what's going on at work—they fear missing either a threat or an opportunity when not connected. For many, it will soon be intolerable to commute from home to office or office to Starbucks or Starbucks to home without checking their e-mail. Employees also feel a higher level of importance—perhaps accurately—if management deems it necessary to equip them for always-on communication. In short: users have a high crisis.

But there are other participants with greater clout in the decision-making process. CEOs want their forces to be ever more mobile and BlackBerry fits the bill as one tool to enable greater mobility. CEOs also have a crisis.

CIOs and CFOs take their marching orders from above so if mobile e-mail is deemed a necessity these two have the responsibility to make it happen. Translation: Crisis. For their part, service providers are desperately looking for additional revenue, so mobile e-mail gets high attention from their sales forces. Again: Crisis.

Total Perceived Pain of Adoption

Contrary to popular belief, the folks at the heart of the decision-making process are not so price sensitive about enterprise mobile e-mail at $45 a month. But there are definitely critical factors that vendors of this technology must address in order to convince major corporate decision makers.

What do CIOs and CFOs care about if not price?

SECURITY
and
SUPPORT

What's the best way to address these considerations? Reference accounts. CIOs and CFOs want to know that they will not get fired because they messed up on security or support. They want to get on the line with a bunch of their peers who have already cleared these hurdles. They want testimonials to reduce their total perceived pain of adoption. They can get them from Research in Motion.

FUTURE WINNER 3: BUSINESS INTELLIGENCE SOFTWARE

"Traditional reporting vendors like Actuate, Information Builders, and Business Objects are 'basking in the scalability wars' as user companies look to bring BI to the masses."

—Wayne Eckerson, The Data Warehousing Institute, quoted in *Computer World*, January 31, 2005

Crisis: High
TPPA: Fairly High

Business Intelligence software promises to enable organizations to collect all their data and information and knowledge for intelligent queries that optimize critical business decision making.

Background

IDC defines Business Intelligence tools in three segments: Online analytical processing—which includes querying, reporting, and analysis—data mining, and preconfigured single-package data mart/data mining.

The ultimate goal of BI is to provide individuals with insight and visibility into operations and to empower users to make effective decisions that improve the performance of their business. That's both a decent proposition and an easy sell. Consequently, organizations have never before been so eager to adopt BI technology. However, a lack of alignment between people, processes, and technology has led to many misguided BI deployments.

Many organizations currently find themselves—as many did after the enterprise resource planning or ERP waves—with underutilized and overpurchased technology that does not provide a competitive advantage. Several challenges that continue to plague modern BI systems include keeping applications properly synchronized to support mixtures of components from different versions of the same product lines, integrating divergent systems following mergers and acquisitions, and integrating the different technologies and interfaces operated by customers, distributors, and suppliers.

As BI software increasingly becomes mainstream, BI vendors are attempting to address the above issues by reevaluating their mix of analysis and reporting tools in an effort to improve scalability and performance. Vendors are updating reporting tools with stronger and more flexible data-analysis capabilities and boosting the performance of online analytical processing tools. Although enabling BI inside organizations can appear simple, behind the scenes is a complex network of people, processes, and systems. Synchronizing these three areas is critical to improving the success of BI initiatives.

Market

> *"In 2004, the [Business Intelligence tools] market experienced a better-than-expected performance, and preliminary research shows the market growing by 9.5 percent to reach $4.25 billion in worldwide software revenue. IDC forecasts a 2004–2009 compound annual growth rate of 6.0 percent"*
>
> —IDC, 2005

There are many ways to define the business intelligence market, and IDC's is one of them. We think IDC's sense of wide-ranging maturity misses the broader trends supporting faster growth rates across time. In short, business continues to get more competitive and global and generally brutal. Organizations will continue to seek tools that enable better decision making. Business Intelligence software provides answers.

That said, unlike rocket-shot technologies that experience meteoric penetration rates, a major cultural shift is required of users when they take on the painful transformation to the software. The result: it's more likely there will be a steady stream of converts instead of the total potential audience making the transformation in one specific year. As a result, business intelligence software has a high likelihood of above average growth for a long period of time.

The Upshot

Business Intelligence (BI) software is an antidote for complexity—the greatest curse in the enterprise. The last antidote—so-called enterprise resource planning, or ERP systems—failed to parse through massive databases to help make faster, better informed decisions. BI is more compelling in the power that it imparts to its users. Want answers based on decades of sales and performance metrics? Sure! The problem is, the TPPA of such solutions is still high. First, the data quality that is drawn up into business intelligence systems is still suspect. Second, very few

CEOs know how to query a database. But just imagine that critical point when it becomes clear that such queries permit managers to become that much more knowledgeable in all the decisions that they make.

Crisis

Today's business world is becoming infinitely more complex, and modern companies generally have a large number of applications that take care of running the business. Such application diversity initially wasn't much of a problem because the applications were meant to automate self-sufficient independent functions, so there was initially little concern that the result of this diversity would be a mismatched collection of "stove-piped" applications rather than a unified network of linked systems.

But the resulting complexity is perhaps the single largest crisis facing enterprises in relationship to technology. The crisis: how to integrate and make sense of all this data.

An initial attempt to solve the integration of these diverse systems came from vendors such as SAP, Oracle, and PeopleSoft that began marketing ERP systems. All ERP applications are "preintegrated" because they come from the same vendor at the same time, so adopting an ERP system theoretically eliminates the need for heavy investment in application integration. But ERP systems failed to make the enterprise fully intelligent at all. There still exists a dire need to link all sorts of information from all sorts of databases to all sorts of users in all sorts of ad hoc circumstances.

The next step in the evolution of application integration strategies comes in the form of business intelligence software that enables businesses to see and use large amounts of complex data. The crisis for integrated data on a real time basis grows each year and those demands come from a far wider audience than ever before. Everyone seems to want integrated data to prove their positions or defend their positions from attack.

It's likely business intelligence software will experience steady 15 percent compound annual sales growth over the next ten years as opposed to the rocket-shot flat panel display television is likely to have in a very short period of time.

TPPA

Why the relatively sluggish epiphany? Users' total perceived pain of adoption is high. Very few of the people who want what business intelligence software provides have ever interfaced with a database for more than a short period and, therefore, have little clue as to how powerful these tools can be. But the culture will steadily shift and education about the power of the tool will expand steadily over the next decade at which point many CEOs will carry a pocket device they can use to query databases all over the world in an integrated, near real-time fashion. The migration from here to there will be powerful. Vendors that make business intelligence simple to the end user will dominate those focused on additional features.

> *"Hyperion is introducing a drag-and-drop, wizard-driven dashboard development tool to its Hyperion Performance Suite 8.3, enabling business users* without programming skills *to develop personalized dashboards."*
>
> —*eWEEK*, March 7, 2005

That's my emphasis in the above quote. It's all about lowering TPPA.

FUTURE WINNER 4: SATELLITE RADIO

> *"Just a blink after the newly emergent titans of radio—Clear Channel Communications, Infinity Broadcasting, and the like— were being accused of scrubbing diversity from radio and drowning listeners in wall-to-wall commercials, the new medium of satellite radio is fast emerging as an alternative."*
>
> —Lorne Manly, *The New York Times*, April 5, 2005

Crisis: Moderate/Growing
TPPA: Low

Satellite radio promises to bring users a far greater amount of well-organized, crisp, dependable, and commercial-free content than that provided by traditional radio.

Background

Many people flip between stations searching for music without commercials while on the way to work or when running errands. But we all know the familiar experience of having the signal break up and fade into static. Most standard radio signals only travel about thirty or forty miles from their source. The promise of satellite radio is wide ranging, abundant, and commercial-free, near CD-quality music from a signal more than twenty-two thousand miles away.

Satellite radio has been over a decade in the making. In 1992, the U.S. Federal Communications Commission (FCC) allocated a spectrum for nationwide broadcasting of satellite-based Digital Audio Radio Service, or DARS. The FCC granted two licenses in 1997. CD Radio—now Sirius Satellite Radio—and American Mobile Radio—now XM Satellite Radio—paid more than $80 million each to use space in the allotted spectrum for digital satellite transmission.

Car manufacturers have been installing satellite radio receivers in some models for a few years, and General Motors has invested about $100 million in XM Satellite Radio since it began installing XM receivers in 2001. Honda, Hyundai, and Toyota have also signed agreements to use XM radios in their cars, giving XM a large advantage in that distribution channel. More recently, several models of portable satellite radio receivers were introduced to the market from a variety of electronics companies. In total, the automakers that have signed on for factory-installation agreements manufacture 86 percent of cars sold.

Although each company offering satellite radio has a slightly different plan for its system, there are several key components present in both, including satellites, ground repeaters, and radio receivers. XM Radio uses two Boeing HS 702 satellites, appropriately dubbed Rock and Roll. The first XM satellite, Rock, was launched on March 18, 2001, with Roll following on May 8. XM-3 was launched on February 28, 2005, as the company needs to replace both Rock and Roll by 2008. XM Radio's signal contains over one hundred channels of digital audio. In urban areas, where buildings can block out the satellite signal, XM's broadcasting system is supplemented by ground transmitters. The Sirius system consists of three satellites that were launched on November 30, 2000. Sirius currently offers both car radios and home entertainment systems.

Each XM receiver contains a proprietary chipset, consisting of two integrated circuits designed by STMicroelectronics. XM has also partnered with Pioneer, Alpine, Clarion, Delphi Delco, Sony, and Motorola to manufacture XM radios. XM and Sirius both charge a $12.95 monthly fee. An extra radio can be sponsored on a family plan for an added $7 per month. As of early 2005, 13 percent of XM users took this family option. XM Satellite has teamed with AOL to provide a total of two hundred channels on the Internet—seventy from XM and one hundred and thirty from AOL—for $12.95 per month. Sirius offers online services as well. The radio itself costs approximately $150 plus installation. Auto manufacturers tied with XM typically give away three months of service to buyers, and manufacturers working with Sirius tend to include six to twelve months free.

Market

When he forecasted the market in a February 17, 2005, report entitled "Initiating Coverage: Satellite Radio Sector," UBS's Lucas Binder projected that total subscribers in North America would grow from 4.4 million to 55 million in ten years. He's since upped that estimate to 62.7 million, and recently reported expecting 39.7 million users by 2010. For perspective, there are 200 million registered cars and light trucks in the United States and an additional 10 million heavy trucks and boats. There are approximately 16 to 17 million new cars sold each year in the United States.

The Upshot

Satellite radio is radio—only better. There's far more content—as in one hundred-plus channels—including plenty of sports from around the United States. Across time, the word *satellite* will likely disappear from the phrase "satellite radio," as there will be little reason for such a distinction. The crisis will quickly build. On the total perceived pain of adoption front, satellite radio is not so much a new technology but more of an upgrade with near zero education required. Across time, the sales channel will be dominated by factory-installed satellite radios in a growing number of the sixteen to seventeen million cars sold each year in the United States. The total perceived pain of adoption will drop even further as the requirement for a trip to Best Buy will disappear as the service comes bundled with the car itself.

Crisis?

Folks want *more* choice and satellite radio provides that choice. Folks want *fewer* commercials and satellite ameliorates much of the pain. Inside a few years, satellite radio will *be* radio. I expect it to be standard in quick order.

Total Perceived Pain of Adoption?

Buying a $150 per year contract at the point of purchasing a $30,000 car is painless. Just as important, anyone who understands how a radio works will learn satellite radio in under a minute. Much like flat panel display, there already exists an enormous market that's familiar with the basics of the product. A great combination.

EIGHT

ONE FUTURE LOSER: RFID

"The [Auto ID] labs will continue to conduct R & D initiatives in pursuit of the Auto-ID Center's original vision, which includes, in our opinion, a quite futuristic view of RFID and associated technologies, promoting the notion of a ubiquitous network that will make it possible for computers to identify any object anywhere in the world instantly. While this long-term notion is intriguing and seductive, it appears to be based more on theory than practical implementation at this point. Adoption of new technologies or new ways to implement existing technologies never seem to materialize as fast as industry 'experts' project, and we do not see why RFID would be any different from the established precedent."

—Adam Frisch and Jason Kupferberg, UBS, November 2003

Crisis: Low and Concentrated
TPPA: High and Diffuse

RFID promises to save expenses through the food chain through greater efficiency, better inventory management, lower personnel expense, lower theft, as well as better stocking of popular products to reduce revenue loss that occurs when sales walk to other providers.

Background

Automatic identification is the broad term given to a host of technologies that are used to help machines identify objects, including bar codes, smart cards, voice recognition, some biometric technologies, optical character recognition, and radio frequency identification (RFID).

RFID is a generic term for technologies that use radio waves to identify people or objects. There are several forms of RFID, but the most common usage is to store a serial number on a microchip that is attached to an antenna, enabling the chip to transmit information to a reader that converts the radio waves into digital information that can be passed on to computers.

RFID technology has been routinely spotted since the 1970s, but until recently it was too expensive to be considered practical for most commercial applications. The major advantage RFID has over bar codes is that is it not a "line-of-sight" technology requiring that a scanner can "see" the bar code to read it—RFID tags can be read as long as they are within range of the reader.

Despite these advantages, it's unlikely that RFID will completely replace bar codes, because bar codes are relatively inexpensive and effective for most tasks. Instead, it's more likely that RFID and bar codes will coexist for many years.

RFID is currently being used for any number of tasks, from tracking pets and livestock to triggering equipment down oil wells and automating tollbooths on highways through very successful applications like E-ZPass. The most common applications are payment systems, access control, and asset tracking.

Potential

> "ABI's report, entitled RFID Service, saw the majority of the 137 companies that sought to meet RFID mandates last year investing only modestly in RFID technologies. . . . ABI's research found

that many started small and allocated budgets far below the $2–$3m that many industry analysts had predicted."

—*Computergram*, January 6, 2005

"There is the potential for benefits in recall management, but that alone might not build a case for adoption."

—Kevin Murphy, *Computergram*

IDC expects the production of RFID tags to climb seven-fold between 2005 and 2008. I don't think the growth will be nearly so dramatic. I expect the breadth of deployment through global food chains will be surprisingly limited. RFID isn't likely to hit widespread mainstream use for years. We may find ourselves still waiting twenty years from now for this global RFID ecosystem to evolve from niche applications and trials to something pervasive.

My own personal hope? I hope that none of the highway authorities in the states connecting New York to Maine read this piece. They might either cancel the fantastic RFID-enabled E-ZPass toll program or just blot out my own specific records, thus placing me forever in a long queue of cars filled with all the other scoundrels fishing for quarters.

The Upshot

There are two very big problems with RFID. First, the winners are mainly limited to one node in the extensive global technology food chain, and those benefits are still fairly unclear. Though publicity has been high, we are barely into the first stages of trials of the technology. There will be a lot to learn from these initial modest deployments before further commitments are made. What's more, the costs are theoretically absorbed by those who gain no benefit for their trouble and who will likely resist participating.

RFID GROWTH DRIVERS AND BARRIERS TO ADOPTION

GROWTH DRIVERS	BARRIERS TO ADOPTION
Need for supply chain efficiencies (i.e., cut labor costs, better inventory tracking)	Tag and reader costs
High costs of shrinkage and "out of the stock" conditions	Accuracy of technology
Opportunity to reengineer supply chain business processes	Lack of standards for interoperability
Compliance deadlines set by large enterprises (Wal-Mart, Department of Defense)	Privacy concerns

Source: UBS

Crisis

I do *not* think there is a legitimate widespread crisis in RFID land beyond a handful of limited trials aimed to see if there are some niche crises that RFID can address. The *high* profile projects are moving along much more slowly than the average person would suspect. In other words, there's a low crisis level. The only benefits seem to be concentrated at one specific node in the vast global food chain.

The few very high profile attempts to make something of RFID have very little game plan and very little idea of the difference this technology would or would not make. The general media thinks this is a great story and seems to swirl around the promise of RFID routinely. It's a whole lot of noise. My take? We are very early on in an uncertain take rate. The attempt to jam de facto standards into the world and the obsessive media focus on getting the price on RFID tags down and standards set are overdone. The optimistic expectations that RFID will be *big* are forced. RFID has not been thought through to conclusion amid the media parade.

The reason that no one has much of a crisis is that bar coding—

where standard codes were developed in the early 1970s—works pretty darn well in those places that expensive RFID projects are being contemplated or induced. It is generally good enough to know that a product is a certain type of product—a case of soda. The incremental information that it a *specific* case of soda is overkill. What to do with all this data? Who knows?

The first bar code patent was awarded in 1952: U.S. Patent #2,612,994 to Joseph Woodland and Bernard Silver.

1962   Bernard Silver dies before the first commercial use occurs.
1966   First bar-code application commercially attempted.
1970   UGPIC code standards set—became UPC standard.
1974   The first UPC bar-code swipe—ten packs of Wrigley's Gum at Marsh's supermarket in Troy, Ohio.
1992   Joseph Woodland awarded the 1992 National Medal of Technology.

The point: new cultural and social rituals require time.

Slap and Ship

> *"U.S. retailer Wal-Mart is using its unmatched market power to compel RFID implementations at its suppliers. Wal-Mart put a January 2005 deadline on its top 100 suppliers to start using RFID and EPC. These suppliers were obliged to join EPC global. This has resulted in what have been known as 'slap and ship' RFID projects at these companies. They slap a tag on a pallet of goods before it leaves the warehouse, an act of corporate compliance to keep Wal-Mart happy."*
>
> *—Computergram*

As suggested above, Wal-Mart kindly told its top suppliers in 2003 that they might want to kindly adopt an RFID program by January 1, 2005, if they wished to keep doing business. Nearly everyone of the top one hundred Wal-Mart suppliers met the bare minimum decree to use RFID when shipping to Wal-Mart, but most just slapped a tag on a box and didn't track it themselves, but let Wal-Mart feel good that they complied because they're Wal-Mart and you don't want to annoy Wal-Mart. There's no way that supplier number 79 will spend $10 to $20 million on a full-fledged RFID project.

LIMITED DISTRIBUTION CENTERS AND PRODUCT LINES

Wal-Mart's RFID program is actually more or less limited to four distribution centers and to a handful of product lines.

Does that make sense? A slow trial? You bet. Technology trials are a *good thing*, and lots can be learned at a reasonably low price—particularly if you force much of the cost burden onto other people! But is the full-fledged RFID supply chain inches away?

No.

In 1992, my wife, Kelly, and I moved from New York City to the Mainline outside Philadelphia, and I could hear again. The noise was gone. It was so quiet I could hear birds and, I swear, I could hear the trees barking and the grass growing. With RFID, I just hear noise. I don't like noise. I like hearing the real "signals." The signal-to-noise ratio is very low amid the RFID chaos.

Radio frequency identification has been an ongoing topic during my seven-year association with the TTI Vanguard think tank, and when it first came up in conversation and someone suggested that the technology could be used to ensure certain expensive wines would remain properly chilled during long transport, I thought . . .

"Geez, that doesn't sound like a big opportunity."?

When I heard that RFID could be used to ensure that beef hadn't spoiled prior to being sold, I wondered *just exactly* how much beef did spoil prior to sale and then how much of that was sold anyway, and I thought . . .

"Geez, maybe it's a good thing I don't eat too much red meat."

When I heard Intel, which is always on the prowl for a new market, was passing on RFID tags I thought . . .

"Geez, if we're hoping that these tags go to a penny, then you would need to own about 100 percent of a market about twenty thousand times as large in unit terms as the PC market Intel currently serves in order to gross an equivalent revenue figure. And geez, could the margins be anywhere near as amazing as those Intel enjoys in processors?"

With those demons in my brain, I was still assured that the RFID opportunity was really big and that, no, RFID was not just a fancy technical way of reaping largely the same benefits that bar coding already provided. I was assured repeatedly that I just didn't get it.

"But, geez, the math just doesn't make much sense to me."

Today, the entire concept still doesn't make a lot of sense to me. A reminder:

A 10x technological change in no way automatically translates into a 10x change in user experience. While technologists will create 10x technological changes over and over and over again, it's the user experience that *really* matters.

As amazing as it is that a technology can recognize that one pallet in a warehouse is missing without a physical check system, it is unclear that the move from bar codes to RFID tags constitutes a tenfold change to end users.

COMPARISON OF BAR-CODING AND RFID TECHNOLOGIES

	BAR CODES	RFID
Transmission technology	Optical	Electromagnetic
Typical data volume	1–100 bytes	128–8,000 bytes
Data modification possible?	No	Yes
Position of reader relative to item	Visual contact	Line of sight unnecessary
Reading distance	Several meters (within line of sight)	Centimeters to meters (depends on transmission frequency & tag type)
Environmental susceptibility	Significant (dirt, intense light, etc.)	Very limited
Anticollision*	No	Yes

Source: Accenture *and* UBS

*Anticollision refers to the capacity of the system—while bar-code scanners cannot read more than one bar code (item) at a time, RFID readers have the capability (if enabled by supporting software applications whose development is still in the incubation stage) to read multiple RFID tags while avoiding issues of "collision" among numerous tags which may enter the reader's electromagnetic field simultaneously.

THE REAL END USER

> "*Payback will be slow to come, however. Most survey respondents using or testing the technology estimate it will take two or more years to see a return on investment of RFID technology. In fact, one in five say they aren't sure they will ever see a return.*"
>
> —Beth Bachelor, *Information Week,* March 28, 2005

If Delta Airlines starts tagging all of our bags, who's the end user? You? Me? Or Delta?

Fliers may want better service, but it's not abundantly clear what service the tag itself provides to the airline passenger. Airlines have contemplated tagging bags for their own cost reduction purposes as opposed to creating a better user experience. The end user is Delta. At least the end user *was* Delta—their program is on hold as their current financial predicament seems to be more important than lost bags, not to mention hot meals and free newspapers.

With RFID, the "user" so far tends to be the entity one step away from the food chain's termination point. *Termination point* is a fancy way of saying "you and me."

In retail, the purported sweet spot of RFID, if you or I are the final node of the food chain, it's Wal-Mart, Target, and Tesco that seem most interested in benefiting from what RFID may provide. That's pretty much it—even one node up the food chain, there are surprisingly few companies bubbling over with enthusiasm about RFID. Of course, Wal-Mart's partners will be compliant if they must. But that group thinks bar coding does pretty darn well, is effective and cheap, and that the benefits of a major disruptive RFID change management process are vague, modest, nonquantifiable, and costly.

And here's the major problem with RFID from an adoption standpoint: The person paying for the technological infrastructure is different from the "end users"—Wal-Mart, Target, and Tesco—benefiting from the technology.

I asked John Dillon, CEO of Navis—a logistics software company focusing on shipping ports around the world—just how many of the world's fifty largest ports used RFID today. His best guess? Seventy-five percent have had a flirtation with RFID paid for by a third party, but none of them use RFID for anything other than an occasional high value container carrying gold or bullion. John doesn't expect much more to happen since the folks with the crisis—Wal-Mart?—are not the same folks paying for the infrastructure.

According to IT Services provider Unisys, the number of handoffs—or nodes—in the average international shipment is . . .

17

In one sense an infatuated RFID bull might say, *Wow! Think of all the inefficiency to solve! What an opportunity!*

But it would also—and more importantly—be the case that the Change Function would therefore need to be applied not just at one or two points but to all seventeen points on a typical international shipment in order for RFID to match up with its glorious image. An RFID-enabled supply chain is only as strong as its weakest link.

And it's not going to happen the way RFID's boosters would like. Why? Because there is little crisis. Who has the incentive to do this? You could force Procter & Gamble to slap and ship one node prior to getting to a Wal-Mart distribution center, but what about the first sixteen nodes of the chain? There are six hundred thousand warehouses in the United States alone, and most of them don't even use bar codes. Is it reasonable to think they want to leapfrog to RFID if they haven't yet had a high enough crisis to move to bar coding somewhere during the past forty years? Probably not.

WAL-MART'S CRISIS?

> *"Mayberry RFID. Radio-frequency identification technology will also become a household concept in 2005 as retailer Wal-Mart plans to have RFID systems installed in as many as 450 stores by year's end. It won't happen without support from suppliers, some of which are going along grudgingly."*
>
> —Scot Petersen, *eWEEK*, January 3, 2005

Why Wal-Mart? What's the crisis? Of all the companies out there, doesn't Wal-Mart already have one of the most efficient supply chains? Perhaps—Tesco gets higher marks for effectively using technology—but Wal-Mart's historical use of technology suggests why they have a "crisis"—in a somewhat positive sense—and many folks do not.

Wal-Mart has proven extremely adept at taking advantage of significant changes in technology to the detriment of competitors. Whether RFID offers Wal-Mart an explicit positive return-on-investment may be a secondary point. The real issue: whether RFID creates a discernable competitive advantage. Wal-Mart deploys technology for a living, and they do it pretty well. If there is a technology with a ray of hope, Wal-Mart will pursue it. To not do so would be to break with corporate culture.

So Wal-Mart had a crisis and ordered its top one hundred suppliers to spend gobs of money to become RFID-compliant by the end of 2004 or else. Wal-Mart's idea was to use its size to drive standards, but in the context of a world of suppliers with seventeen-node chains, suddenly Wal-Mart seems only so big.

Can Wal-Mart stimulate a crisis at Target? It seems so. How about beyond Target? Not really. Until standards are set and a large percentage of goods are moving more effectively with tags on them, other retailers will choose to be fast-followers as best possible and more likely will hope the standards process gets mucked up along the way. Let Wal-Mart and the myriad RFID vendors figure it out and then get synced after the kinks are worked out.

A SECURITY CRISIS?

In the wake of the September 11 tragedy in New York and other awful tragedies around the world, many folks thought that "security" could be the inducement toward RFID adoption. Might governments mandate the implementation of RFID systems to check items entering a nation? If so, who is going to pay for it?

If everyone has a low level crisis in relation to security, everyone has a high total perceived pain of adoption at nearly every point along the seventeen nodes to fund it. What about the folks shipping the goods? Are they going to pay for the readers at every node? No. What about the node operators at the XYZ terminal in Genoa, Italy? Nope.

Less than 5 percent of containers arriving in the United States are currently physically inspected. Is it possible in the aftermath of catastrophe that regulation might be adopted to mandate some form of inspection? Yes—but it hasn't happened yet.

IT SUPPLIER CRISIS?

Amid a global maturing of the enterprise IT space, is it surprising to see a wide variety of IT suppliers waving the RFID flag? Not at all. They are desperate for growth.

But that doesn't mean anything particularly special is actually occurring, and these vendor announcements can't jawbone a market into existence. Still, in the event that RFID business picks up, they will be able to claim that they *have had an RFID practice for XYZ years . . .* when they go on sales calls.

> *Just 'cuz folks keep talking about something over and over and over doesn't mean its gonna happen! Remember Y2K?*

Interested parties include those companies making readers and transponders and chips—the hardware—plus consultants and middleware providers and Indian outsourcers and ERP folks and even Intel is now finally entering the game after initially snubbing the entire space. Intel is suggesting they can optimize their chips with SAP solutions for what RFID requires but it seems to us to be a real low priority at Intel.

The company in the most interesting position is probably Verisign—named by EPCglobal to manage the Object Name Service root that assigns and tracks the numbers associated with individual tags. Verisign is doing what it can to lower the total perceived pain of adoption by driving the EPC2 standard by launching a free network for development and testing.

> *"EPC, Electronic Product Code, is the numbering system that will be used by most radio frequency identification tags. It is similar to bar-code numbering, but is it big enough to enable products to be identified by unit, not just by type."*
>
> —Kevin Murphy, *Computergram*

It's a big job. Without much of a crisis.

Total Perceived Pain of Adoption

Short answer: *very* high. Reducing the cost of the RFID tags placed on pallets is a necessary ingredient for RFID adoption, but in no way, shape, or form is it sufficient. The pain only begins with RFID tag costs.

One industry contact put it this way:

> *The industry isn't structured to make [RFID] work on a global scale—he who pays doesn't necessarily win—a bad business proposition if ever there was one. Also, they "forgot to mention" the TCO [total cost of ownership]—sure the tag is cheap—but all the readers in all the repair depots? And then the "overhead" checking tags against boxes for fraud, maintaining databases and systems changes—that nobody will do for free—etc, etc, etc.*

The largest pain of all is process change management—shifting from a deeply imbedded technology that works to a new technology that doesn't work yet. Process change management is terrifying to executives, so if the costs of RFID are borne by folks with nothing to gain, why would anyone lift a finger if they could avoid doing so?

> *"If you're piloting an RFID network in your warehouse in a controlled test environment, everything will probably work fine. But in the real world, when you have, for example, 75 cases of beer on a pallet rolling by a reader at 10 mph, problems arise."*
>
> —Ephraim Schwartz, *Infoworld,* November 26, 2004

TAG COSTS

Technologists often struggle to separate necessary conditions for technology adoption with sufficient conditions for technology adoption: In the case of RFID, it may be a necessary condition to bring RFID tag prices down a lot, and it may be a necessary condition to develop standards everyone can build around and write code to, but they are not *sufficient* conditions.

There is no market-clearing price for an ugly shirt!

Like nearly every build-it-and-they-will-come technology, there has been a Moore's Law fairy tale spreading about price points of tags. The thinking goes like this: after the tags—and there are several varieties of passive and active tags to consider—get to low enough price points, the market will take off. In typical fashion, this fairy tale has been slow to die. When forecasters pick magic price point number one and it fails to stimulate the market, magic price point number two is created, and so on.

Price is but 10 percent of the total perceived pain of adoption.

Low price points are typically a necessary though not a sufficient condition for adoption of a new technology. What's the ultimate motivation to put RFID tags on boxes of razor blades in mass quantities?

"Because occasionally a pallet of razors contains 195 boxes of blades as opposed to 196?"

I don't think so.

The most convincing argument for a large-scale RFID rollout is based on smart shelves that notice when all the Gillette blades are gone so someone could be yelled at to *Get some more of those blades on the shelf!* This heads off two potential problems: if Gillette-loaded shelves sell out and the shelf is empty and someone buys something else—Gillette's problem—or the customers leave and don't buy anything and go to another store—the retailer's problem. So, Gillette decided to start a Smart Shelf project. The Smart Shelf would know what had been taken and how much remained. When inventory ran low, the Smart Shelf could demand a refill and would therefore, theoretically, always be properly stocked—thus reducing the possibility of missing out on sales when a customer fails to find what they are set to buy. But the Smart Shelf project seems to be dead in its tracks after being canceled about two years back, ironically during the frenzy of RFID. The vision of what might work some day vastly diverged from the reality of the day.

> *"I believe we're years away—maybe three or five years—from tagging individual items, especially low-value items."*
>
> —Brian Matthews of Verisign, quoted in *Datamonitor*, January, 24, 2005

The reality is more like these items may not be tagged for ten or twenty years.

THE WHOLE PRIVACY THING

> *"I'm a big proponent of weighing potential risks versus potential rewards. So this push for RFID in so many sensitive areas of people's lives confuses me because the risks are so high and the rewards seem so minimal."*
>
> —Jim Rapoza, *eWEEK*, March 7, 2005

Now, with all of this going on, users were also hearing that RFID tags were to be tremendously feared. While we suspect RFID's failure will lie at the hands of other more pertinent issues—like *who has an incentive to participate?*—concerns about privacy have now and again gotten in the way. In reality, much of the privacy fear spread among those who might buy products with tags affixed was ridiculously exaggerated. People imagined that others would know the contents in their wallet or purse. Our former household IT guy, "Tech Geek," convinced my wife that the U.S. government had already inserted chips into all currency in order to track us all. (My confidence in him was shot after that comment.) Nonetheless, privacy concerns raised the total *perceived* pain of adoption.

- Folks such as the Electronic Frontier Foundation and others have warned with regard to consumer privacy issues.
- Benetton halted its plans a couple years ago to tag all its items, as consumer groups feared that these devices might somehow track them.
- A smart shelf project was pulled when it was found that consumers were unknowingly being videotaped in order to see reaction and usage.
- A grade school in California failed to tell parents that student badges contained RFID technology allowing tracking.
- The United States is planning to drop RFID in passports and is soon expected to demand the same of visitors.

Are some of the privacy concerns wildly out of touch with reality? Yep. Does that mean they are unimportant in the process of making RFID a widespread mainstream technology? Nope. Just one more of the many hurdles RFID will need to overcome.

NINE

THREE MORE FUTURE LOSERS

There are more losers out on the technology landscape than winners. That's to be expected when 90 percent of what is offered up fails to connect commercially. Below I highlight just three more future losers.

FUTURE LOSER 2: FIBER TO THE PREMISE

"There is no requirement for FTTP to gain share in access lines. Right now, it's not on our front burner. There's no sense of urgency to take FTTP for a compelling product offering."

—Randall Stephenson, COO of SBC, 2005

Crisis
Service Provider: High
Residential User: Low

TPPA
Service Provider: Significant
Residential User: Low

Fiber-to-the-Premise—any combination of fiber to either the home, curb, or node—promises to offer consumers faster pipes for sending and receiving data in their homes as well as all-in-one "triple play" bundles of voice, video, and data.

Background

Fiber-to-the-Premise—FTTP—refers to a telecommunications system based on fiber-optic cables for the delivery of broadband service to homes and businesses. Fiber-to-the-Curb and Fiber-to-the-Home fall under the umbrella of Fiber-to-the-Premise, with the distinction being exactly how close the fiber gets to the end user. The closer to the user, the more expensive is the build. FTTP gained momentum insofar as press announcements are concerned in May 2003 when BellSouth, SBC, and Verizon announced that they had adopted a common set of technical specifications for the delivery of FTTP in the United States.

The initial announcement was followed by a joint RFP in June 2003 that was issued to selected vendors for FTTP equipment. At around $1,500 per installed fiber line, a program involving over one hundred million lines could easily drive a return to profitability for many equipment vendors that had been chastened in the telecom and Internet collapses.

FTTP has been touted as the "Next Big Thing" for broadband in the United States, and KMI Research forecasts that the total FTTP market for equipment, cable, and apparatus will grow to $3.2 billion in 2009. Despite high hopes for FTTP, deployment and implementation of the network have been rife with challenges and hesitations. The most aggressive proponent of running fiber past residences is Verizon. Their grand target for the end of 2005 was a mere two million homes.

Although Verizon is the most publicly bullish on FTTP, they failed to reach their goal of one million homes connected to their FTTP network by the end of 2004 and the technology rollout continues to lag their initial goals. Deployment difficulties aside, FTTP is expensive to deploy. Verizon is committed to undertaking a copper overbuild project—replacing their installed copper lines with fiber—while BellSouth and SBC are, for the most part, limiting fiber runs to new homes and businesses. But even with the high costs, there is a consideration that FTTP could provide a longer-term competitive advantage in that the Bells may not be forced to share these networks as they must with copper.

Potential

It's likely that build-outs will be at least somewhat close to stated schedules in 2005 and 2006. But take rates from those homes passed will be disappointing, leading to an awkward moment twelve to eighteen months down the road when the telcos will need to adjust their plans, offer deeper discounts, or both, and come back to investors with an explanation and a new game plan.

The Upshot

Occasionally very impressive technology developments are deployed even when the user does not experience anything remotely close to a 10x change. The key question to consider with Fiber-to-the-Premise technology is "whose crisis is it, anyway?" The answer: the telcos. It's unlikely the telcos will roll out at the pace implied in brash public relation announcements since a boom in cap expenditures would likely lead to a vicious downward share-price spiral. Additionally, it's doubtful many consumers would want to endure a fresh install for what today is largely me-too capability.

Crisis

The telcos in the United States have a crisis in that cable companies finally seem poised to introduce voice services through their pipes thus creating a so-called triple play of voice, video, and data. Comcast, for example, plans to enter forty markets with voice service by mid-2006 and is targeting about eight million subscribers. Users are already getting these services, and would be happy to save some money and consolidate into one bill. That said, none of this in itself is a 10x user experience change.

TPPA

The telcos will roll FTTP out on a tactical basis so as to match their cable competitors location by location but will be averse to breaking the bank on a service users aren't screaming for—shareholders will freak if telcos hike their capital expenditures too much.

FUTURE LOSER 3: THE ENTERTAINMENT PC

> *"Everyone's been waiting for the great convergence product. [But] . . . people are not ready to replace their televisions with their PCs."*
>
> —Toni Duboise, senior analyst, ARS Inc., as quoted on
> news.com, September 3, 2002

Crisis: Low
TPPA: High

The promise of the entertainment PC is not exactly clear. Whatever it involves, it's supposed to be better than what we have.

Background

The "digital home" has become a catch phrase for all sorts of vague ideas. The problem is, not everyone is downloading movies through Bit Torrent or loading photos to Flickr or has heard of Wikipedia or has bought a personal video recorder or eliminated all wires from their home entertainment systems. Very few people are bursting for the advantages provided by the mythical digital home.

A minority of people has purchased a flat panel display, a good percentage of homes have broadband, but only a small number are bothering to network their broadband. Few folks have yet to throw out their

CDs in favor of an MP3 lifestyle, but many are tossing out videos. Few people have an Internet refrigerator or like using PhotoShop. Few are watching IPTV, and WiMax has a long way to go before it provides benefits to the digital home. Few people have mastered personal storage manipulation to the degree they most certainly will during the next five to ten years.

As a core element in the digital home, the entertainment PC has barely happened at all. Designed first and foremost for flawless multimedia playback and secondly for playing games, the ultimate promise of the entertainment PC is to change the way users think about photos, music, and movies. Instead of having a shelf full of photo albums, racks of CDs, and a case of DVDs, the entertainment PC allows you to keep everything in one central place. When connected to the Internet, it also can be used to download and purchase music or movies. Although multimedia on a PC has always been a dream for computer manufacturers, the industry's first real attempt at an entertainment PC wasn't until the latter half of the 1990s when Gateway introduced its Destination line of computers. These machines combined a large screen, fast CPU, DVD player, and TV tuner, but Gateway's Destination line failed to catch on because the systems were very complicated to use.

During the next few years, many manufacturers toyed with entertainment PCs, but the next real progress occurred in mid-2002 when Microsoft shipped the first edition of Windows Media Center. Windows Media Center is Windows XP with an additional user interface bundled into the operating system that allows users to watch TV, record TV programs to a hard drive, burn DVDs, view photos, and listen to digital music—all from a single remote.

Perhaps the biggest benefit of using an entertainment PC is the elimination of extra components, because entertainment PCs can store large volumes of digital video and audio while replacing TiVo, CD players, DVD players, MP3 players, and radios with a single black box that simplifies the cable mess. Despite these potential benefits, the entertainment PC is not resonating, given the current needs of consumers.

Potential

Timing is everything. Sociologically, we are in the midst of a digital revolution. The demographic waves are incredibly important. The number of people who grew up in digital environments where computers were standard issue as opposed to somewhat special—Digital Natives—will keep mounting. The spread of Digital Natives will convince those among us who remain terrified of technology—analogists—that they better "get it" soon or face social ostracism and risk economic extinction in the workplace. Nevertheless, even with these important demographic forces lurking, it is doubtful the digital home or the entertainment PC will happen at a pace fast enough to satisfy technology companies or Wall Street.

The Upshot

The "digital home" is a catch phrase—as opposed to a technology—chasing a crisis. Until folks bemoan the fact that their DVD player connects to the TV with cables, and the TV to the cable box with—egads!—more cables, having a centralized entertainment hub in the home is just not critical. Sure, consumer electronics makers and even enterprise-focused technology companies are in crisis—seeking new revenue streams by introducing a plethora of cool "digital home solutions," like the entertainment PC, but consumers aren't there yet. In ten years? Sure, but now, we'll live with the wires that are conveniently hidden behind wall units and media stands anyway. In the meantime, any technology that helps analogists get comfortable with manipulating "content" in any sort of way in their home, office, mobile, and over the Internet is a great thing. The world will look very different ten years from now as society's TPPA in regards to manipulating and sending and receiving all sorts of data plummets.

Crisis?

Those with the crisis here are unmistakably the PC vendors and anyone in the PC food chain. Does anyone really think the world is waiting for a meta-controller of all other media controllers? Products that come with names like the Enterprise PC or enabling technologies such as Intel's ViiV—pronounced Vive but for one reason or another not simply spelled *Vive*—promise us another remote control for our lives which may be the single least desired additional device on the planet. Suggestion: if they have to give us our tenth remote control, please color it bright yellow or fluorescent—I suspect Apple has patented the illuminating white—so we don't have to pick up nine remotes to find the one that the machine demands we subordinate ourselves to.

TPPA?

I could write an entire book about the total perceived pain of adopting an entertainment PC—but I won't. The upshot is that what the PC ecosystem finds "easy to use" is quite bizarre to the average human who clearly has not been consulted or observed. One simple question: how many people are willing to spend a weekend studying a user guide to eventually understand all the complex "simplicity" in a feature-overloaded monstrosity? No one. The developers will say, "But it's really easy," and the potential user will say, "No it isn't," to which the developer responds, "You, consumer, are wrong! It's easy!" That's a bad way to run a business.

FUTURE LOSER 4: WIMAX

"I think WiMax is going to be a disruptive technology that's going to change the way we think of mobile connectivity. Hopefully, toward the end of 2005 or 2006 you're going to see massive commercial rollouts of this capability."

—Intel's Craig Barrett, *The Online Reporter*, March 5, 2005

Crisis: Low
TPPA: High

The promise of WiMax is to provide much faster wireless transport speeds over a much greater physical distance than is possible today. The use of WiMax technology promises to be highly flexible for a large number of potential applications.

Background

Worldwide Interoperability for Microwave Access—WiMax—refers to a broadband wireless solution that is based on the IEEE 802.16 standard. In practical terms, WiMax claims to offer wireless high-speed Internet connections to homes and businesses in a radius of up to thirty-one miles without a direct line of sight to a base station. This means that WiMax could potentially serve areas that currently have no broadband Internet access since phone and cable companies are unwilling to run necessary wires to those remote locations. WiMax might also cover entire metropolitan areas, allowing true wireless mobility as opposed to the "hot spot" hopping currently necessitated by Wi-Fi.

The main benefits of WiMax are that wireless access is less expensive than cable or DSL and easier to extend to suburban and rural areas. As standards for the technology are determined, the hope is that customer premises equipment costs, or CPE, will drop from about $800 today to $300 to $400 in 2006 or 2007, according to Mobile Competency. R.I. Meta Group is even more bullish, expecting WiMax CPE to drop to $70 by 2007.

The current 802.16 standard was approved in June 2004, and amendment 802.16e—the most advanced, mobile, and potentially versatile of all the WiMax standards—was completed in early 2006. Korea's telecom industry has also developed its own standard, WiBro. In late 2004, Intel and LG Electronics agreed on interoperability between WiBro and WiMax.

Despite the progress thus far, WiMax still faces significant challenges, the biggest of which is answering the question of who will pay for the technology. Will companies set up WiMax transmitters and then require users to pay for access? Or might cities pay to have WiMax base stations set up in key areas for business and commerce and then allow free usage or wholesale the network to service providers?

WiMax is simultaneously an opportunity and a threat to a large number of parties. Not only does it pose a threat to providers of DSL and cable-modem service, its potential to incorporate VoIP technology—which offers the ability to make phonecalls over the Internet—means it also threatens long-distance carriers.

Intel is the key driver behind WiMax. The company intends to WiMax-enable Centrino laptop processors in the next two to three years. If WiMax is actually incorporated in every laptop by 2008, setting up WiMax base stations may provide a great opportunity. In the meantime, big-name telecommunications service providers still view WiMax as merely one of many technologies to be considered.

> *"The reality is that WiMax has been hugely overhyped."*
>
> —*The Economist*, January 29, 2005

Potential

WiMax has so many *possible* applications that the lack of focus may cause a failure to make any single substantive impression on the world. Is it fixed broadband like MMDS—802.16.2004? Is it for wireless back haul? Or is it a 3G competitor? A curb-to-home for triple play? Or for rural access? For service providers? Or for governments?

Will Starbucks upgrade Wi-Fi to WiMax? As the question of what WiMax is *technologically* is settled, the bigger question of what WiMax is *practically* will dominate this conversation. The lack of focus combined with many competing technologies suggests WiMax is unlikely to be the runaway winner its largest supporters suggest.

The Upshot

In addition to assuming that the technology works flawlessly and that standards are imminent, fans of WiMax center their case on a few variables. They may assume that Intel—in its public support for WiMax—will drive adoption; that carriers will support it; and importantly, that users are dying for an upgrade from Wi-Fi. Though Intel has shown success in tapping into latent demand by selling into enterprises or consumers, this time it is selling to telcos. Are telcos interested in deploying WiMax, a technology that might cannibalize their existing revenue streams or serve as a reminder that their bids on 3G networks in 2000 remain the biggest blemish on their report cards? Not likely. More importantly, while users may one day demand a better, faster high-speed service, at this time, crisis is low. We're still adjusting to using our laptops at Starbucks. WiMax as an earth-shattering change over the next three years is unlikely. I see only niche applications for rural high-speed and perhaps some businesses replacing their expensive T-1 lines over time.

Crisis

End users of all sorts are just beginning to have a crisis in relationship to accessible wireless broadband to connect their laptops to the Internet or their corporate network wherever they may choose. Some would like us to believe that this steadily emerging crisis is equivalent to wanting a specific technology called WiMax. But users don't care how they get online. Suppliers throughout the technology ecosystem will find only a few niche instances where WiMax is the clear-cut answer to their crises. It's important to remember that users have crisis in relation to the service a technology provides as opposed to a crisis in relation to the technology itself.

TPPA

The list comprising the total perceived pain of adoption for WiMax is so long and so multitiered—from users to service providers to numerous equipment providers—that it seems impossible to know where to start or end. Equipment providers with a large stake in selling next generation 3G wireless equipment would be considered insane after all these years of waiting to sell their somewhat expensive gear to abandon course and support the sale of relatively inexpensive WiMax gear! Those benefiting from high bandwidth wireline services such as T-1s are unlikely to get really excited about displacing those expensive services for the relative cheapness of WiMax. End users will want their laptops to work with the expanding cloud based on standard Wi-Fi technology as opposed to prestandard WiMax technology that would require a swap out of their current laptop. Any time we encounter a technology with such a large number of questions to answer about adoption we add several years to the commonly ascribed forecasts.

TEN

THE CHANGE FUNCTION IN ACTION

To truly appreciate the power of the Change Function, it needs to be seen in the real world—not just on the printed page. The following two case studies focused on two very cool companies. Their coolness is not because of their technology—which is cool nonetheless—but because they are immersed in a user-centric culture.

SALESFORCE.COM

> *"The tremendous pressure to sell manifested itself in customer dis-satisfaction."*
>
> —Bruce Cleveland, a Siebel exec in the 1990s,
> *Information Week,* April 4, 2005

As I sit here at Borders in October 2005, Salesforce.com has yet to become a household name. In another year or two, it may well be, as more and more folks determine they better darn well use some form of customer relationship management software as provided by someone such as—or particularly—Salesforce.com. And if it happens that Salesforce.com becomes widely known—even with its dot-com suffix—it will be because of actual success with customers. A hype-laden Super Bowl ad won't do the trick. The company believes deeply that nearly every new brand will be created from the bottom up.

In 1999, I began loosely tracking Salesforce.com as just another soon-to-be-failed application service provider when my friend Heather Bellini—a software analyst at UBS—had the company's management in to chat with us about nine months prior to their initial public offering. I was biased against anything classified as an application service provider at that time so I didn't expect much. But everything changed for me the moment they answered one specific question I asked. I even remember where I was sitting.

To set the context: the proposition of application service providers was that enterprises would take the applications they had spent millions on and hand them over to a stranger—to them, the application service providers (ASPs) themselves. For years, I ranked this among the dumbest ideas I'd ever heard. CIOs and CFOs don't just hand things over—especially if they're tricky and they stand a chance of getting fired if the whole thing goes awry. At that point, no ASP had any substantive reference accounts, and reference accounts are about the only way to reduce fear of firing among CIOs and their ilk.

So I asked, "What is the key ingredient above all else that will make you successful? Somehow building deep trust with clients?"

Their response: "They choose us because we are codesigning and codeveloping the product in a very iterative fashion with small groups of salespeople who are responsible for actually using the software. We don't sell top down to CEO/CFO/CIOs. So our utilization is really high because users codesign what they really want as opposed to what management dictates."

That was my ah-ha moment.

Since so few people were even speaking in a user-centric fashion back in 2002, I was frankly taken aback. And I have been exploring the company ever since. Salesforce.com lives in the Change Function model—they live and breathe the end user. While most people wandering about the technology ecosystem today are wise enough to at least pay lip service to "user experience" and "simplify" and will even hire an anthropologist or two, very few really know what it is to live the words they have learned to say. Salesforce.com is an actual living breathing champion of user experience.

The Change Function says that users change habits—which might include adopting a new technology—if and only if the pain of their current situation is greater than their total perceived pain of making the switch to a proposed solution for their pain.

Way back in the 1990s, Tom Siebel of Siebel Systems helped a niche application become mainstream. It was called customer relationship management software, or CRM. Best known for its use by the sales teams of companies both large and small, CRM is actually useful for anyone in any part of any business that might have a customer—in other words, everyone!

At a small dinner hosted by Goldman Sachs in 1998, Tom Siebel told me and about five other investors that the number one challenge at Siebel would be to educate end users after their companies had bought their Siebel product. The second challenge, incidentally, was to keep the number of expensive sports cars in the parking lot at Siebel down and keep employees thinking instead about the business as opposed to their now affordable high-priced toys. He said all this while we were at Aureole, on the Upper East Side of Manhattan, eating very tall desserts.

There's an important point in there: first, Siebel sold a lot of product. *Then* they dealt with education. Given the framework of the Change Function, such a scenario seems unlikely: why were folks willing to spend tens of millions of dollars on new software when the vendor itself was worried about the problems in teaching end users how to use it? Because in the 1990s the "crisis" with respect to CRM software was at the CEO/CFO/CIO level in most organizations. The crisis was that everyone *knew* that you were *supposed* to buy this stuff to track business and if you didn't you were obviously a dinosaur and certainly the board would wake up and fire you.

That was during the Internet Bubble. By the way, that same day in 1998, Tom Siebel had been at a private CEO round table. The *big* topic had been Internet Strategies and by the end of the day it was clear to Tom that absolutely no one had any idea what it was all about. The result? If those CEOs could muster up just one Power Point slide about their Internet Strategy they could keep the board mystified and investors at bay since those folks didn't know what an Internet Strategy was either.

That was 1998. So buying $10 million of Siebel software to spread across the organization made sense for CFO/CEO/CIO types, even if training was a gigantic question mark and even if they found utilization stunk! It was all worth those *minor* risks. Just tell your people that if they don't comply with the system their sales leads will be curtailed or they will just be fired. That will force them to comply!

The problem for Siebel was that after the Bubble, the old crisis of appearing to "get" the Internet gave way to a new crisis that required CEO/CFO/CIOs to spend judiciously on projects with a positive return-on-investment. It quickly became much harder to sell $10 million of software at a clip.

In the late 1990s, companies were experiencing very low utilization rates with Siebel and other CRM systems because the education issue was never solved *and* many employees felt that management was playing Big Brother and so they decided *not* to participate until they were threatened with dismissal.

> *"What a waste! What does management know about managing my accounts? Nothing! Who is senior management to buy software about my client tracking without asking me what I think first!"*

Utilization rates were typically anywhere from 15 to 35 percent. The dissatisfaction was highly publicized. That's a bad long-term technology adoption strategy: Assume actual users will use a new technology when threatened with dismissal.

> *"Even Siebel's own research showed that as recently as [2004], 41 percent of its customers felt they hadn't gotten the desired results from their Siebel deployments."*
>
> —Tony Kontzer, *Information Week*, April 4, 2005

In stark contrast to Siebel, Salesforce.com spoke my language—a language that would work in the 2000s. Do people *really* care about buying Web-based CRM software? Nah! Ask a salesperson if they care about where their applications are located. They don't. Ask them if they care about its effectiveness and whether it was *their* choice or management's decision. They do!

But there's a hitch: Salesforce.com still has to overcome—along with every other vendor of software geared toward the enterprise—the stigma of being, well, a vendor of software geared toward the enterprise. The animosity in the relationship of software buyer to software seller is pretty palpable. Those selling the software are compelled to complete an epic struggle in order to close every single deal—you eat what you kill, the saying goes, and buyers are truly fearful of getting stuck with a monolithic expensive monstrosity that everything else in the network must conform with, be designed around, be written to.

While enterprises were addicted to the big license purchase in the 1990s, Salesforce.com and other ASPs can address new buyers because they're able to say, "Why don't you just run a pilot to see if you like it? If you don't, you can terminate it." There's no epic sales struggle. No hooks. No five hundred-seat minimums. No one to push back against. What an adjustment.

Salesforce.com's user culture goes much further than merely saying "cancel when you wish." "Major changes" in its software occur two to three times a year and they make minor changes pretty much every week. Where other companies craft a major release once every two years, Salesforce.com intentionally keeps its vision almost exclusively inside six months. Since the business model doesn't require five hundred-seat minimum deals, Salesforce.com is quite happy that their average account today has but seventeen seats.

Providers of monolithic solutions must somehow decide *today* what will sell two years hence when the product is locked down and ready to ship. And it better be different and much better or clients will suggest it is simply an update of the old release in which case they are entitled to it *for free* as part of their expensive maintenance agreement. It darn well better be good. No such drama at Salesforce.com. At Salesforce.com the product is never locked down. Features are never finished. Constant iteration is the mode of operation.

Where do they get the ideas for their changes? From their users, in the form of complaints and suggestions, as well as by studying customer activity itself.

Step 1: Salesforce.com translates customer complaints and requests into possible new features.

Step 2: They also watch the activity in the applications—which they can do since Salesforce.com hosts them. They can see what is and isn't used.

Step 3: They formulate new features and iterate current features based on this information.

Step 4: They test new features with real live clients who might come in to talk for a while about it. In return, Salesforce.com personnel might say thanks tons and give their clients a $10 Starbucks gift card.

Step 5: Design and development gets more feedback.

Step 6: Salesforce.com iterates until they are set to deploy.

Step 7: They ask their users for permission to include any changes—functionality is never forced.

Step 8: Then they start over again, repeating a fairly straightforward low-risk, community-enhancing process.

As Salesforce.com makes clear their responsiveness to customer complaints and feature suggestions, you might imagine that more customers will take the time to make their own suggestions, which will further help in codesign and codevelopment. Across time, users will feel like part of the community in that eBay community sorta way. When you have a "community" as opposed to a bunch of users or clients trapped in a monolithic inflexible software prison, those community members actually buy in and spread the gospel and the effectiveness of each iteration becomes all the more certain. Why bother complaining to a package software vendor when the next release is a gazillion years away anyway?

CODESIGN AND CODEVELOPMENT

What do the designers and developers inside Salesforce.com think about working alongside actual users in such an iterative fashion? After all, a core hypothesis of this book is that technologists like to be left alone to build cool technologies and have the users line up for their creations. Isn't codesign an infringement upon this freedom?

Parker Harris, EVP of technology at Salesforce.com, woke me up to an even greater desire of technologists. It goes like this: technologists want their creations to be actually used! Technologists live to change the world through their work, and that feeling of working on something that never gets out into the world totally stinks. Since new code is predicated on specific user input, Salesforce.com nearly assures that whatever you as a designer or developer are working on will have a great chance of making it into the world—and it's likely to get into the world in a very short period of time.

Salesforce.com believes in hiring the most incredible designers and developers they can find and then, as Parker puts it, breaking down the barriers between them and the users. In their experience, they've found that constant iteration isn't just liked by the users; the designers and developers like it as well.

Another advantage of the Salesforce.com model: it helps its own people work together far more effectively than others do. How so? Well, a major problem in innovative companies is avoiding defensiveness in design and development and support. Since Salesforce.com is continually iterating its product, there is never the one, single, big moment when the monolithic software product is released to the market and the whole organization holds its breath. In those scenarios, everyone preps a game plan of how to blame someone else if the customers don't like the massive once-every-two-years release. If they like it, the storm passes, but if they don't, the finger pointing begins. Since everyone knows how that internal blame game works, you might suspect that the teamwork during the creation and development process might suffer *juuuusssstttt* a little as everyone preps his or her excuses.

But Salesforce.com just keeps iterating. There is no major drama. The best way to reduce the tumult of heightened defensiveness is to avoid a process that stimulates it. A translation for the software industry:

Avoid major releases!

When interviewing CEO Marc Benioff, something came to mind. In some ways, the entire Salesforce.com culture is set in such a way that it nearly *has* to be user centric. "Iterate, iterate, iterate" means no *big* software releases to inspire internal fear of a looming disaster. As such, the attention remains on the customer as opposed to ongoing preparation for the *eventual* inquisition regarding who's to blame for the *big* disaster when customers vote that they didn't like the *big* software release. There is an additional side effect: there are rarely big failures at Salesforce.com because there are rarely *big* situations. Life at Salesforce.com is a series of small steps. As such, the room for cancerous internal defensiveness or poisonous department versus department positioning is dramatically reduced. I asked Marc what his biggest failures were and he said specific hiring decisions here and there along the way. These are important day-to-day tactical miscues as opposed to massive strategic sagas. Nothing is big. Brilliant.

Customer relationship management software will soon get hot—*really* hot. Ten years from now, anyone at a cocktail party confessing that they don't employ CRM software in their daily corporate routines will feel like a full-fledged idiot. The entire global market today might be in the neighborhood of six million seats, but ten years from now that market might be ten times what it is now. Salesforce.com is just setting out on what will be a very long journey.

At this point, it's worth mentioning two minor experiences from at a recent Salesforce.com event at the Moscone Center in San Francisco that were atypical. I have every intention of using Salesforce.com software at my new entity Coburn Ventures, and so I made the trip out from the East Coast to get the lay of the land.

After spending fifteen minutes with a technical type in a demo center, I entered some basic information about Coburn Ventures and our possible interest in Salesforce.com's software. Inside three minutes, I had an e-mail on my BlackBerry from a real live person—that was number one—and we set a time to talk the next day. This was great human follow-up right when I was open to following up. Better yet, the Salesforce.com rep ended up being very soft in the sell process—he actually seemed to understand that buying his software in the next two weeks was not at the top of the priority list at Coburn Ventures. This unusual, nonanimosity-generating interaction was an excellent example of highly effective user-centric selling.

I also asked to see a service rep on site at the Moscone Center since I had a bevy of questions, both for investment purposes for Coburn Ventures as well as for this book. After ensuring I would get a boxed vegetarian lunch, a woman in support walked with me for ten minutes looking for the head of service in my specific geographic region, who we learned was named Joe. She did this all quite—and here's the punch line—happily. That was number two.

That sounds run-of-the-mill or should be, perhaps, but having attended such events routinely for more than a decade, the handoffs are usually and surprisingly so quick and abusive and you are lucky to find one person along a parade of handoffs that seems to think a user-group meeting is for, well, the users—and potential users. The unstated law in most companies regarding conduct at a user-group event is the following: if you happen to encounter a small business user, run away as fast as you can. Big clients matter. Small ones don't. Salesforce.com employees must have missed that memo. At most user-group events, "under ten employees" types tend to be efficiently orphaned somewhere in the process.

No quick abusive handoffs. Joe (happily) answered my questions, all the while knowing I was an under-ten-employee prospect—although he didn't know I would be writing a book describing the quality of the vegetarian boxed lunch, which was indeed excellent.

All this should be good news to Salesforce.com CEO, Marc Benioff.

Salesforce.com's *barrier to entry* is its culture of user-centricity, which leads to codeveloping and codesign and selling to actual users and customer-complaint links, and happy people walking under-ten-employee prospects to their destinations, as well as rapid growth. Salesforce.com had approximately 351,000 users as of October 2005, while last decade's monolithic winner had 3.4 million. Is it a good bet that will all change? Yep.

REACTRIX

One of my partners, Pam, and I had been grinding it out on the roads of Silicon Valley one day not so long ago when we pulled up to Reactrix's headquarters in one of the many office parks that dot Redwood City. We had a 3:00 P.M. appointment with CEO Mike Ribero and his team, and I had only a few moments before I fully realized a major problem with our plan.

Lunch. More to the point, the lack of lunch. We called Mike Ribero's assistant Claudia to let them know we would be fifteen minutes late. We didn't tell them we'd be spending that time scouring their office complex for food. But the one café on campus closed up at 2:30; and we were demoralized—and hungry—when we showed up at Reactrix at 3:15.

You know what Claudia asked as we sat in reception for Mike? "Have you had lunch? I could order you something? What would you like?" I found this amazing. Perhaps the hunger and desperation showed just a tad on our faces.

Claudia dialed up someone and, fifteen minutes later, sandwiches that would rival those found in the box lunches at Salesforce.com's user-group event showed up.

When an assistant asks visitors at 3:15 if they have had lunch, two things might cross the visitor's mind . . .

*Do anything to poach this savant for our own team no matter
the breach of proper etiquette!*

or

*We have just gotten a fantastic glimpse of the user-oriented
culture of Reactrix at work.*

I am not the first to think that I might be able to come to an ultimate investment decision about a company in the first two minutes after entering their headquarters. In the case of Reactrix—a private company—I regret they didn't come ask me to open up my checkbook when they raised their venture capital funding.

That's all well and good, but I should probably move on to what Reactrix does other than making sure its guests are well fed before meeting with company management. In short, they do alternative, new media advertising.

In August 2005, Tony Perkins held his now annual AlwaysOn Summit at Stanford and asked me to moderate CEO pitch sessions in which CEOs of private companies have five minutes and five slides to let the audience in on what they do and why it will become such a fantastic business. Judges then vote and provide their feedback. Mike Ribero presented Reactrix.

Mike was the only CEO who used part of his time to demo his product. He did so for two good reasons. First, without seeing the Reactrix system in action, it's very hard to "get" what it does. Second, Mike Ribero is extremely user centric, so doing a presentation without us—his audience—getting to see what he does is antithetical.

A side note: he had zero to gain by making his presentation at all, since he already has as much money as he needs for his young company. He didn't have to impress us. "Impressing" is supplier centric and self absorbed. "Serving" is user centric.

Here is the part where I will try to describe what Reactrix does. If it's at all possible, I'd suggest first setting the book down and going over to the Sony Metreon Complex at the corner of Fourth and Mission streets in San Francisco so you can experience it. Go ahead. I'll wait.

. . . dum-dah-dum-dum . . .

. . . Okay . . . so . . . you're back . . .

You saw the Reactrix system—an approximately five-feet by eight-feet image on the floor of the Metreon between tickets and concessions? And you saw that the infrared image shone from a projection system in the ceiling down to the floor and you noticed that the images represented a variety of advertisers like JAL Airways, Adidas, and Orville Redenbacher popcorn.

What I hope you really noticed is that the image on the floor *reacted*—hence Reactrix—to your movements as you entered the advertisement space by standing on the Reactrix system. Perhaps you saw the Adidas image that showed a soccer ball rolling around an image of a pseudo-soccer pitch. Did you try to kick the ball? If you did, it reacted and moved in the intended direction as if you had *actually* kicked it. Whoa.

Cool technology. Adidas, you say? You remembered that, did you?

Or you might have noticed that if you stepped on any of the popcorn kernels on the Orville Redenbacher ad, the image of the popcorn popped and made a popping sound. You might have noticed older people watching their kids have fun. And you might have noticed a whole lot of people smiling. Reactrix is fully engaging as a medium.

Cool technology. And they'll remember Orville Redenbacher. This is a technology I "got" a whole lot better after I experienced it.

In this age, traditional advertising media like newspaper, radio, and television are all losing sway (to be kind) or steadily collapsing (to be more direct). Pay-per-click online advertising from Google and Yahoo threatens newspapers. Satellite radio is undermining radio ads. Personal video recorders will steadily shrink the number of times TV ads are actually viewed. There is a large appetite for new forms of advertising. I am writing this just a week after *The New York Times* missed Wall Street's earnings expectations for the fifth quarter in a row. So, the advertising industry's crisis means things like Reactrix's engaging systems are getting a good look by all.

But even if we agree that this technology is cool and we agree that traditional media forms are under pressure, isn't it possible that Reactrix could just be a fad? Possibly. So I asked everyone I have met and talked with at Reactrix: "Isn't this just a fad like, say, 3D movies?"

And they all say that it is about the content not the technology—but I kept just seeing the technology. Company founder, Stanford grad, and Google alumnus Matt Bell offered up that they are able to check actual usage and "really good" content keeps users engaged three to six times as long as average content. If it was just about the technology, then the Adidas soccer match or the popping corn wouldn't hold people's fancy any longer than something else. Bad material won't cut it. Good stuff will make a big difference.

So how does Reactrix help make the content special? That's at least partially the responsibility of Janis Nakano, who heads *experience design* for the company. In most places they call this *content creation* or such, but the phrase *experience design* makes it clear that the goal is not to create clever, sophisticated, impressive, award-winning industry-adored masterpieces but rather content that engages users.

Janis says that experience design is happening throughout the retail industry because competition is just way too intense. Buyers can now buy anywhere but their experience of that "buying" can vary considerably. Providers must offer users a better experience in buying the same product in their "place" as opposed to someone else's. If they don't, they won't be around very long.

Long ago, I asked former consumer-goods consultant Adam Devito about industries that thought about user experience as opposed to just trying to supply a cool product, and he pointed me to his old space of consumer-goods products. Go talk to Starbucks or Procter & Gamble, he said, because this user stuff is all *they* talk about. It may just be that I've been hanging out in the technology world for too long, but when assistants are asking us about lunch at 3:15 or companies have experience design groups, I am venturing into a new world. And I like it.

But Reactrix still has to create content. Or cocreate it. Just like Salesforce.com, Reactrix works with its customers to create the ads. In fact, the Reactrix model is to train third-party developers who will create their own content using Reactrix resources as guides. Eventually, with the development of simpler and simpler tools, Reactrix may get out of actual experience design itself. At least that's the plan.

For now, Reactrix leverages its own in-house MD in Neurologic Science as part of the process in creating engaging material. It's hard—outside of iconic Silicon Valley design firm IDEO—to find such a talent that combines significant academic grounding with substantive industry experience. If you find such talent, you may want to grab some for yourself—since it takes about ten or twenty years to develop. Reactrix isn't a company that hires an anthropologist as an expert witness to support preordained conclusions. Reactrix believes that content will be the difference between a fad and a sustaining medium so they plug cultural expertise into the process. Huge difference.

But these experience design gurus don't operate in a windowless world. There is plenty of data to feed the process since Reactrix is a public event. Users can be observed across a network of systems—not in a simulated lab but rather in the real world. Reactrix people, who are trained in observing, can do the whole "learn-relearn" thing ad infinitum. They can iterate to their hearts' content.

The creation of new content costs only $5000 a copy since Reactrix's clients use their own intellectual property—images, brands, etc. So we don't have an issue that each Adidas soccer-field spot rivals a major motion picture in costs thus creating the fear of massive disasters. It is only $5000. But getting the content as engaging as possible can be tricky! Reactrix makes their money through access to the network of one hundred-plus systems around the country.

So . . . since content is componentized inside the development process, Reactrix is able to iterate the material on the fly any time they wish even after the content is out in public by just tweaking the code in those smaller components without a major rewrite. Changes can be made when they see when real humans interact with the ad in the real world as opposed to "testing" in an artificial lab or focus group and then praying once it is birthed into the real world.

This sounds strikingly similar to the iterative process at Salesforce .com. Iteration in the real world.

Is codesign formulaic? Not really. Codesign is iterative, and iterative means that the compass of one's overriding beliefs about what end users will react to positively dominates over a religious protocol to do something one way and one way only. As Matt Bell put it to me, "Process can eradicate soul." The overriding belief at Reactrix is that for advertising to work it must be engaging and it becomes most engaging when you cede control to the consumer who decides how and in what way to interact with it.

The creation process must render content that has such a quality. It isn't so much that process per se is a bad thing but when process dislocates companies from their value system—their soul—a lot of mediocre results are sure to follow. When we run our own internal investment meetings at Coburn Ventures, we use process only to the point that it aligns with or complements the soul of what we do. When meetings turn energy-less we know something has gone awry.

Reactrix shares a common trait of all great user-centric entities. They are unsatisfied. Matt Bell is at least humble if not unimpressed with what Reactrix systems have provided to users to date. He realizes there is so much more the technology can enable that new content could reside on top of. He realizes that as a live entity Reactrix must hit short-term goals in order to get to the long term. He is a businessman.

He is also a technologist who wants to change the world. He's been working in the background on new surfaces for his systems beyond floors—such as walls and tabletops. He thinks of additional technologies to layer in, such as directed sound that would only be heard if one was on the Reactrix display but not otherwise—cool. And he thinks of new applications with an ultimate goal of improving education. Matt yearns to get his technology out in the world to make a difference, and he doesn't feel like he has even yet started. He is dissatisfied. The culture in Reactrix can sense this, but instead of dwelling on how little they think they have accomplished in the grand scheme of things, they are driven by the opportunity to create an even greater difference through technology in the years to come. They are excited. I see this quality in every great technology company I have ever encountered.

ELEVEN

TEN SETS OF QUESTIONS

"For seven and a half million years, Deep Thought computed and calculated, and in the end announced that the answer was in fact Forty-two—and so another, even bigger, computer had to be built to find out what the actual question was."

—Douglas Adams, *The Restaurant at the End of the Universe*

Sitting on my porch up in Maine on a splendid—though nippy—October Saturday, life couldn't be better. I'm listening to the new U2 album in iPodworld as I open my ThinkPad with a fresh task—to make *The Change Function* a relevant tool for you, the reader.

It's kind of fun being a tech historian and analyzing failed technologies, but Michael Schrage implied that these failures exist for a purpose—to learn from them. I agree.

Here's where I'm supposed to say, "Fear not . . . here I present the eight easy steps for employing the pearls and perils of the Change Function into your daily life and all will be well." Sorry. Eight-easy-steps books are never easy and are usually abandoned pretty quickly. The purpose of the Change Function is to uncover greater truth, understanding, and better assessment of organizations, decisions, products, managements, and outcomes. Each of the following questions starts with the Change Function as its reference point.

Adoption of new technology = f (perceived crisis, total perceived pain of adoption)

"Somewhere along the way the promise of the technological revolution to make our lives easier is not being delivered."

—Gerald Kleisterlee, CEO Philips, as quoted in the
Financial Times, September 14, 2004

Tech may *happen,* but it isn't *working* so well. Again, it isn't about the core technology working. The failure of the vast bulk of technology innovations to hit business targets isn't driven by the failure of the core technology to do what it promised. Rather, these technologies are not geared toward their customers. Customers who may need these technologies are—in sharp contrast—terrified of the solutions provided! So technology isn't *working.*

The following questions, grouped into ten categories, are meant to help you understand precisely what a company, an entrepreneur, or a management team is doing and how well they're positioned to succeed with their new products and ideas knowing the odds are stacked heavily stacked against them.

**For an entrepreneur: will great technology become
a great business?**
For a manager: can we become serially successful?
**For an investor: will this product, management, entrepreneur,
or company succeed?**

There are no perfect answers to these questions, but there are answers that work much better than others. It is a litmus test. What to do with the results is up to you.

GOOD AND BAD ANSWERS

What should we be looking to learn when interrogating management teams?

- Are they stuck in the 1990s view of easy growth?
- Are they stuck in the Moore's Law x Grove's Law world?
- Can they understand the Change Function?
- Do they have a sense for altering user crisis?
- Do they understand total perceived pain of adoption?
- What are their design, development, and marketing processes?
- Can they precisely break down the potential market?
- Can they adapt their thinking?
- Can they materially change their organization?

Generally speaking: answers that focus on either Moore's Law or Grove's Law, such as *We just need to get the price a little lower to take advantage of demand elasticity*, and so on. All those answers are . . . *bad*.

Generally speaking: answers that focus on either digging deep to uncover core crises or identifying a "crisis that can be inspired" and answers that involve reducing the users' total perceived pain of adoption, such as *The user hated version one because the GUI wasn't intuitive enough and we brought in a new team from Apple to fix it*, and so on. All those answers are . . . *good*.

CATEGORY 1: THE CORE SENSE OF SUCCESS

The questions below may seem like pretty basic ones. The point of asking them is to see if the management team that's sitting smack dab in front of you in the physical world is actually living on a different mental planet. Do they think change is *easy*, or do they think change is *hard*? Do they see the introduction of a new product as *change* or as *technology*?

The core issue: many technologists think that cultural change is easy once you've conquered the technology. I don't agree at all. Cultural change is hard.

> *What is your sense of the percent of newly introduced products that will fail??*

A *terrible* answer to this question would be *Between 20 percent and 50 percent of new products succeed.*

It's pretty unlikely than anyone—not even someone from a distant planet—would say, *Above 50 percent succeed.*

Below 20 percent but above 10 percent is heading in a better direction.

A *good* answer is *Less than 10 percent.*

A *great* answer is *Less than 5 percent, but let's get into why we think the odds on our product are 30 percent and what precisely we need to do and the breaks we are counting on in order to turn that 30 percent into reality.*

> *What are the core aspects of successful products today versus, say, 1998?*

This question aims to determine whether management is living in the long-gone "build-it-and-they-will-come" world or the emergent "crisis" versus "total perceived pain of adoption" world.

Terrible answers include *The product feature set was technology, just much more advanced* or *The cost finally got down to the right price point.*

These answers reflect the Moore's Law x Grove's Law obsession and a lack of true insight into potential clients' minds.

A *great* answer might include: *Well, we finally made the product simpler for the user; and, therefore, the potential addressable audience expanded,* or *After layering the product with too many features, we realized that beyond the tech geeks, the specific feature that a larger audience would be drawn to was XYZ, and we altered our marketing program to get that one single message across,* or *Word of mouth finally spread, creating a cri-*

sis among some users that was greater than all the pain they assumed they would need to go through to actually use the product.

> *How would your model differ if we were still in 1998?*

This is a slightly different take on the previous question, and yet again the goal is to understand what world the management team lives in.

A *terrible* answer might be *Not much. The growth rates have been slower the past couple of years, but you never know when a buying frenzy may start, and, since there has been a drought recently, we think [inexplicably] that there is a pent-up demand for new ideas and technologies today.*

A *great* answer might be *Well, back in 1998, users were in a panic to buy technology. After the three bubbles crashed—Y2K, Telecom, and the Internet—there's been a far more skeptical audience that demands that technology work better and be easier to use. Technology is no longer the wunderkind it was, and so we need to better cater to large group of folks that are even more suspicious. The good news is that certain people are more comfortable than ever with technology—especially younger people who grew up with it—and they look at our product much differently. Back in 1998, those people were younger, had fewer dollars and less ability to affect decision making in the enterprise. So here's what I would do specifically if it were 1998.*

CATEGORY 2: TRACKING SUCCESS AND FAILURE

> *"Champions of new inventions display persistence and courage of heroic quality."*
>
> —Donald Schon, *Harvard Business Review*, 1963

This set of questions is oriented toward determining how a management team handles product development and rollout. Do they "get" the low odds of new products clicking in the market? And what are their identifiers of success, beyond mere sales numbers? For example, press mentions and critical acclaim were very important for Microsoft's entry into gaming with Xbox. These markers were even more important than sales since Microsoft was building a multigenerational platform from the start and needed critical acclaim of their system to merely get a chance at round two. They did. Conversely, Nokia failed miserably in rolling out N-Gage handheld gaming on mobile devices. The industry buzz was wretched. For Research-in-Motion's BlackBerry, a key tracker is celebrity placements. Oprah has one, if you didn't know that already.

> *What are the criteria you'll use to decide to pull a product if it isn't working?*

One *great* answer would be *If we get the sense that the user experience is not anywhere as interesting as we thought it would be and we realize we were wrong in our key presumptions and they are not fixable for whatever reason, we would shut down the effort immediately.*

Having generated approximately three hundred reports on investing in technology, I've had my share of real clunkers with regard to presenting ideas to clients. I often had some idea I thought would *really* help our clients get their arms around something. But it didn't. Not even close—just confused them further. When I "get" that an idea *really* doesn't work, I do one basic thing:

Stop!!!

"The first rule of holes: When you're in one, quit digging . . ."
—Molly Ivins, *Austin American-Statesman,* January 1, 1994

There's no better way to find out about the culture of an organization than asking how failure is handled. If failure isn't handled well, the culture resembles a prison. Why a prison? If there is punishment for thinking out of the box and failing, you can bet that most people act in a way that's consistent with the avoidance of punishment. They're in a self-imposed prison that limits their abilities to take risks, make decisions.

In such situations, it's difficult to expect innovation in products or marketing or "new thinking" to emerge easily. And if an organization is slow to change—particularly a technology organization—watch out! They are likely trapped in the Moore's x Grove's Law paradigm and will *not* be early in the revolution.

Another *great* answer would be *We would pull the plug if we didn't get traction in our key early target markets. We would either need to realize our core thesis was faulty or that we need to reconstitute the product for a different type of success. But it would be back to the drawing board. At all costs we need to avoid groupthink after our first decision to go to market has been made. Thomas Jefferson once said, "The moment a person forms a theory, his imagination sees, in every object, only the traits that favor that theory." We need to guard against that.*

This answer reveals a dissection of the market consistent with the *personal experience* nature of the Change Function. It often takes time for an accurate theory to play out, so management must be looking for the subtle clues along the way. Quick exits aren't the necessity, but rather great thinking consistent with the Change Function.

> "*Sometimes we measure things and see that in the short term they actually hurt sales, and we do it anyway.*"
>
> —Jeff Bezos, quoted in *Fast Company* magazine, August 2004

A *terrible* answer: *If we don't hit the sales targets, we'll pull it.*

This answer *seemingly* scores points for cutting the effort loose, but an answer should include a core understanding of the real thesis—and failure thereof—that the product represents.

> *What will account for your product not being successful?*
> *What considerations do you have about the product?*

These are really the same question slightly reassembled that opens two different lines of examination. The first is an analysis of how the core thesis came to be wrong, and the second is checking on the feature, design, or business issues created by the product. The goal is to find out what these guys are *really* fretting about as they walk through their daily lives—whether it's the lack of a QWERTY keyboard or the lack of system management software or a small, independent, software-designer base or worries about the interface.

I remember an interview with Google's Larry Page back in 2003 in which he said—and I paraphrase—*lots of people think our search engine is great, but I think it is pretty horrible.*

I smiled at this *great* answer because, in the grand scheme of things, all these products are horrible as compared to what may be possible. Larry Page probably wasn't referring to Google's ability to return a search in 0.3584 seconds or whatever it is today. More likely, he was thinking of the user that is wondering what key words to use and is frustrated in creating a search that yields no results and plopping through a few screens to pick the "relevant" result only to click and find that it isn't relevant at all—all these things happened when my wife, Kelly, and I searched for a resort in Taos, New Mexico—and in that case, he's right. His service is terrible—and *that* is all an opportunity that maybe Google can solve.

Information technology today is as far away from the man-computer symbiosis nirvana that JCR Licklider dreamed of in 1962 as it was then.

I was exceedingly fortunate to meet an industry legend, Douglas Engelbart, several months ago. Among many things, Engelbart is known for having created the mouse. Is Doug excited about the progress the industry has made as he crusades to make computers usable? No. In fact, he couldn't be more the opposite. Engelbart thinks we have barely budged from the day in 1950 when he consciously began his crusade—but he remains at it fifty-four years later.

A *great* answer would be *We're thrilled with what we have done so far, but in reality, we've barely begun.*

> *What do you most mean when you say* We just need
> to execute?

Ask management or an entrepreneur a question about the readiness of the technology, and they assure you it is set. You ask, *What else is there to do?* and some engineer or software designer by trade who also happens to be CEO of this little enterprise says, *Well, we just need to execute from here.*

Terrible answer. To *any* question!

The appropriate response? *What does that mean? What falls under the confines of "execute"? Do you mean little things like product design and development, manufacturing, marketing, and sales? Just those little details? Or: Under the confines of "execute," what really needs to happen? Could you spend ten or fifteen minutes on that to catch us up with the challenges of actually making some money off this concept?*

It may be that they've just grounded themselves into the same old boring routine of saying things like *we just need to execute* and they are yearning to talk to someone who actually cares. Or they may have no idea what they're talking about, and the *execute* part isn't even fully understood.

I met with tech sage Jaron Lanier recently. He let out a big sigh in response to my comments about *execution* and lamented that so many of the little companies that come to see him—nominally to ask advice about making products and businesses from technology—are so rooted in the Moore's Law x Grove's Law world that they cannot possibly accommodate a different vision. Jaron is thus helpless in helping them. These people see *execute* as a catch phrase for all the parts of business that they hate or are incapable of accomplishing. It's a good idea to find out which type of management you are talking with.

CATEGORY 3: MOORE'S LAW X GROVE'S LAW, PART ONE

"When it came time to begin instructing people about Gypsy, I went straight for the lady with the Royal Typewriter, figuring if I could teach her, it would be clear sailing for the rest."

—Tim Mott, quoted in Thierry Bardini's *Bootstrapping*

These are leading questions. We are on foreign soil so to speak—for all the Moore's Law x Grove's Law folks out there, they will be in their element and you will watch them rummage through many ideas that will reveal their stripes.

> *Forgetting profitability for a moment, can you describe your sense of price elasticity for the product?*

A world in which you can forget profitability and focus purely on *price elasticity,* which is just a fancy way of saying "Moore's Law," is exactly the kind of world that technologists dream of—particularly if the stock market pays them handsomely without all the stuff that needs *executing.*

A *terrible* answer would be *Listen, Bluetooth chipsets are at $25, and we think they just need to get to $5 for this market to take off.*

And you say, *Why $5 exactly?*

And they kind of look at each other and, well, they really have no idea.

Good answers to the question would revolve around price elasticity being a part of a larger picture but not the sole variable of potential success. Sure, they might provide some sense of where they want to get the price of their product—Henry Ford wanted to get the Model-T under five hundred dollars way back when—but they should also discuss the deeper thinking as to what would need to happen socially or culturally in which groups to get liftoff.

Is your product disruptive, and if so, why?

If you ask about disruption, and they talk about functionality, features, and deep technology but never mention the user's experience, you may have a problem on your hands. Putting two and two together, you have just backed into a build-it-and-they-will-come type of management.

Good answers would look at the "disruptive technology" question from the user perspective and downplay technology for technology sake. Something like *Well, technologically—if that is what you mean— there is a major functional improvement here, but* more importantly *we think potential users in certain segments will really experience something remarkable.*

> *"You must design a product that is remarkable enough to attract the early adopters—but is flexible enough and attractive enough that those adopters will have an easy time spreading the idea to the rest of the curve."*
>
> —Seth Godin, *Purple Cow*

CATEGORY 4: MOORE'S LAW × GROVE'S LAW, PART TWO

> *"I changed the law once already, from doubling every other year, in 1975. . . . That cycle could slow down to every four or five years."*
>
> — Gordon Moore, at a July 2002 White House ceremony, during which he received the Presidential Medal of Freedom

After technological considerations are eliminated and you're happy with the price points, what else needs to be done for this product to sell well?

This is a trick question that targets a couple of issues. First, it's another chance to see if this management team is living in the Moore's Law × Grove's Law world.

If they are, they may well say something like the following *terrible* answer: *Well . . . nothing . . . the technology and the price point are all we need to focus on here.*

UGH!

But if instead, they offer up a *great* answer like the following: *That's only the start, and I hope we haven't given the impression that deep technology is all we are working on. We need to get the total perceived pain of adoption down a long way.*

In this case, you may want to say *Have you been reading Pip's blog on Always On or something? Where did you get this goofy total perceived pain of adoption jargon?*

At this point, you want to learn *much* more. What are they actually working on? Why and how much of what they are working on comes from a deep understanding of the potential user? And who is working on these projects and what is their background and pedigree? And so forth.

CATEGORY 5: CRISIS

> *"Management must think of itself not as producing products but as providing customer creating value satisfactions. It must push this idea into every nook and cranny of the organization."*
>
> —Ted Levitt, "Marketing Myopia," *Harvard Business Review*, 1960

These questions dig at the depth of management's understanding of their potential client. It seems simple enough, but experience shows that about 90 percent of the companies out there are far more skilled at figuring out how to develop a Grove's Law disruption than how to develop, market, and sell a disruptive service to their client base.

Technology companies are far more capable of controlling their own core technology development than figuring out how in the world to actually make money off this know-how. Understanding that fact has even turned into a business strategy in and of itself: during the past ten years, Cisco Systems has capitalized on this industry-wide richness in technology development cohabitating with a paltry understanding of clients' needs. Cisco buys much of its R & D. Most of those folks lack Cisco's savvy, skill, and immense customer base. For years, Cisco has been faulted for not developing its own successful R & D—as if *that* was the point of a business.

The real point, rather, is selling services to clients who embrace them. Cisco's competitors are surely jealous of its financial success, but it's also the case that technologists still dominate the technology industry, and not building your own successful R & D is viewed as . . . *just inherently wrong.*

> *"In the case of electronics, the greatest danger which faces the glamorous new companies in this field is not that they do not pay enough attention to research and development but that they pay too much attention to it . . . they are growing up under the illusion that a superior product will sell itself."*
>
> —Ted Levitt, "Marketing Myopia," *Harvard Business Review*, 1960

Either way, it's important that a management team is not beholden to *technology* but, rather, is beholden to serving potential customers. There's a big difference.

What "service" are you selling?

Back in 1960, Harvard's Ted Levitt wrote an article titled "Marketing Myopia." In it he suggested that there really are no "products" but rather every "product" is a service aimed to solve a problem or—in our vernacular—a "crisis." He famously observed that *people don't want quarter-inch drill bits—they want quarter-inch holes.*

Terrible answers to the question above would start with a puzzled look when you even begin to outline the question or the terms that you use. If you suggest that they provide a service, and they are stuck thinking they build a product, then you're off on the wrong foot. If they say *Service? No . . . you see . . . we build widgets.*

Great answers—conversely—would reveal something beyond your own suspicions. If these folks are for real, they live twenty-four-hours-a-day with this question of *What service do we provide?* Whatever that revelation, you will get a sense of exactly how well they know their target audience.

Great answer: *Every person who joins our company reads the Ted Levitt article during orientation.*

In *The Innovator's Solution*, Clayton Christensen provides numerous examples of selling. What service is McDonald's selling with a vanilla milkshake? According to Christensen, they're selling tranquility to parents since it takes little kids so long to actually drink a milkshake. They are quieter as they sip.

Such questions also allow you to see clearly if management deeply understands just what—and whom—they're competing with.

Christensen offers what would constitute a *good* answer: "What is BlackBerry competing against? What gets hired when people need to be productive in small snippets of time and they don't pick up a Black-Berry? They often pick up a wireless phone. Sometimes they pick up a *Wall Street Journal*. From the customer's point of view, these are the BlackBerry's most direct competitors."

> *What is the user crisis you intend to solve?*

The king of *terrible* answers when the topic of "crisis" is raised might look something like this: *Well, I'm not sure about "crisis" really, and we don't provide a service so much as a product . . . and what we need to do is to just get the customer to blah blah blah . . . and then try to execute.*

The words *get, customer,* and *to* don't belong in the same sentence. In reality, they are unfortunately often found right next to each other. At this point, run far away from the coming crime scene. What's so wrong with these words? They are techno-centric and supplier-centric. The custom of managements wanting to *get* customers to *do something* is hopeless in that customers will do what they will do after all the dust settles. We can't "get" anyone to do anything they don't decide to do.

CATEGORY 6: TOTAL PERCEIVED PAIN OF ADOPTION

> *"Just remember: you're not a 'dummy,' no matter what those computer books claim. The real dummies are the people who, though technically expert, couldn't design hardware and software that's usable by normal consumers if their lives depended upon it."*
>
> —Walter Mossberg, *The Wall Street Journal*

The point of all of these questions is to see if a management team understands its customers or even *any* customers. You could easily spend all day on them.

> *What are the top five reasons a user with this "crisis" will not buy this product?*

A *terrible* answer would be *Well, it really just comes down to price.*
Worse: *I'll give you the top five reasons someone with this "crisis" wouldn't buy—price, price, price, price, and price.*

> *What can be done to create the lowest total perceived pain of adoption?*

A *great* answer would be *Well, there are seven specifics we have targeted that need to be reduced in order to reduce the pain of adoption, and price is number six. There are five things we consider more important than price. Would you like to run through the list?*

"Sure, thank you"

> "[Innovators] haven't a clue what new costs their innovations will
> impose on potential users."
>
> —Michael Shrage, MIT's Technology Review, October 1, 2004

> When do you introduce the client into the design process
> and why?
> What internal processes do you use to see your products through
> the eyes of potential customers?

> "The high-tech industry is in denial of a simple fact that every per-
> son with a cell phone or a word processor can clearly see: Our
> computerized tools are hard to use."
>
> —Alan Cooper, The Inmates Are Running the Asylum

Most management teams will come up with a politically correct answer to the above question that sounds like they include clients in the
design process, but really all they're referencing is complaints from the
sales force as to why they couldn't sell the prior version of the product.
Some of those answers may have *actually* represented what some actual
clients complained about. But the goal of the question is something a little more forward thinking. Does management use one of the increasingly popular frameworks—such as the use of Alan Cooper's personas,
which I mention a bit further ahead—that commonly assist companies
in getting at what their customers are *really* like? If so, give them a hundred points.

> Have you introduced value for the client during the past two
> years that many people at first thought against your
> best interests?

This is a question that gets at both the time frame in which results are demanded for "investments" as well as a management's conviction that they understand the essence of their market so well as to push their vision through to reality across years as opposed to months. It doesn't take a lot to have a five-week plan these days, a two- or three-year plans requires enough conviction to push past all the sideline naysayers.

> *"He [Jeff Bezos] has introduced innovations that have measurably hurt Amazon's sales and profits, at least in the short run, but he's always driven by the belief that what's good for the customer will ultimately turn out to be in the company's enlightened self-interest."*
>
> —Alan Deutschman, *Fast Company* magazine, August 2004

CATEGORY 7: CUSTOMIZED WEIGHING MACHINES

> *"One of the things that mark me as an ubergeek power user is my willingness to systematically identify and explore all of these various configuration screens and controls. . . . In the long run, true usability will come when we learn how to write software that needs less configuration, not more."*
>
> —Simson Garfinkel, MIT's *Technology Review*, October 1, 2004

The Change Function
f (Perceived crisis vs. Total perceived pain of adoption)
is really a weighing machine.

If crisis is greater than total perceived pain of adoption, change will happen—technology will be adopted—but if crisis is less than the total perceived pain of adoption, then change will not happen. Every potential customer has a unique weighing machine. Questions to management that see how they see their customers can be incredibly revealing.

> *Who, exactly, is your customer?*

A *terrible* answer would include intense and uncontrolled fumbling about when asked to describe what the *actual* customer looks like and what specifically goes through that customer's head.

A *great* answer might start with *Okay, I am in the persona of the possible user. My name is Rick. I work as an administrator at a hospital. I am twenty-eight years old and. . . .*

> *What does crisis vs. TPPA look like for the first 5 percent of your potential market? What does crisis vs. TPPA look like from 5 percent up to 15 percent?*

Why ask about the potential market in such a way? There are a few reasons. First, management teams tend not to focus too well on their specific target audiences when they initially roll out a product—they often have the idea that everyone *should* use their product. They're disinclined to eliminate even long-shot possibilities from the target market since the desperation to make any sale is particularly high. And that fear leads to a lack of focus. Fear leads to sales scatter, to chasing anything, hoping that something or anything pops up.

It's important to know that management can differentiate and focus. A scattershot approach on the first 5 percent, in particular, is a *really terrible* answer and will typically result in disaster.

After the first 5 percent, the next 10 percent of penetration is another group or groups of users with distinct personas. But there is no 10 percent—let alone 50 percent—if there isn't a 5 percent. Here, again, it's important to gauge management's ability to differentiate and know their market. You shouldn't presume to know the answers. Instead, you want to understand that they can break down the specific market rollout, because if they don't get the first 5 percent in a reasonable order, money will vanish.

> *What will the crisis vs. TPPA look like above that level?*

One need only ask specifically about the first 15 percent—and then merely in general beyond that—for two reasons: first, few ever get anywhere close to penetrating 15 percent of their addressable market, so scattered wishful thinking beyond that is fairly irrelevant. Second, and more to the point, the forces of peer pressure in all varieties tend to kick in above 15 percent, and detailed dissection at that point is probably less relevant and less accurate than with the first 15 percent. That's when *tipping points* kick in.

> *How quickly will you be able to penetrate each layer?*

Questions like this aim to "get" management's sense of likely outcomes and to see the extent that the company has developed a detailed thought process on distribution rather than having gotten hung up solely at the product level.

CATEGORY 8: INTERNAL DESIGN AND DEVELOPMENT

> "Developers looked down on designers because their thinking seemed fuzzy and unstructured, their tastes arbitrary. Designers felt that developers were unimaginative, conservative, and given to rejecting their designs out of hand without trying to find a way to make them work. Because programming was inexplicable to designers, they had no way of assessing a developer's insistence that their designs were unprogrammable."
>
> —Fred Moody, *I Sing the Body Electronic*

Not only do 95 percent of the products emanating from the technology cauldron fail to match expectations, but there are *ugly* cultural factors causing the consistent failure. Worse yet, these cultural factors often function as an impediment to possible change that might make the entire system more effective. It's called humanity.

> *What doesn't work in your R & D, design, and development*
> *process?*

This is a reality check on management's sanity and awareness of how their company works.

A *terrible* answer might show up in the form of *Oh, we just need to execute. It all works pretty well together.*

To which I would offer the open-ended question *Pretty well? . . .* and then I'd shut up and see what they cough up. In a lunch-type setting, this is a good time to look down and busy oneself, making no eye contact whatsoever. They might just cough up something valuable—like the soul of their organization.

> *What steps do you take to eliminate the typical tension that ex-*
> *ists between the design and development teams?*

A *terrible* answer might start *Well, that's just the way it is and not that much can be done—designers and developers will always have creative differences.*

> *What is the most effective element of your hiring process?*
> *Do you use Briggs Myers in the hiring process?*
> *What ongoing training and organization do you use?*

There arc lots of creative ways to hire people, and it's important to make sure that management teams are focused on overcoming possible culture clashes in the design/development process.

Briggs Myers is a personality survey managers use to be better prepared to see the possibilities in which teams may work better or worse together. The mention of Briggs Myers is just a filler to see if there is an attempt to manage the entire design/development process in a holistic fashion.

A *terrible* answer is to find that all these questions merely bore the management team.

[yawn]

A *great* answer is one where you can't shut them up on these topics *and* they use the words, *we iterate,* a number of times along the way and then you shut them up because you still have three more questions to ask and a drive to the airport to catch the last flight.

> *How do you avoid the siren call of tech for tech's sake?*
> *How do you avoid the sales force kitchen-sinking dilemma?*
> *How do you avoid the second system syndrome?*

Camden, Maine-based consultant and writer Tom DeMarco once warned me that the designer/developer relationship isn't the worst of the lot. He is more afraid of a feedback mechanism. Afraid of feedback. Seems wacky, huh?

Specifically, he is afraid of salespeople. Here's how it goes:

The salesperson fails to make a sale.
The sales manager wants to know why.
The salesperson goes back to the customer.
The salesperson asks, "Why, oh, why, oh, why didn't you buy our lovely product?"

The customer when pressed, says, "It didn't have XYZ."
In reality, the customer really just wants to be left alone.
The salesperson tells their sales manager, "I could have sold it to
   them if it had XYZ."

Before long, other salespeople are delivering different messages as to how they *all* could have sold a lot more product if it only had these gazillion more features.

Guess what Tom DeMarco is afraid of: the sales manager reports back to senior management, and senior management tells design and development to load in a gazillion more features. These features are unfortunately off base. The customer-inspired *features* were suggested by buyers who wanted to get salespeople off their backs and out of their offices.

Talk is cheap when a checkbook isn't in sight.

Perhaps the salesperson is just plain ineffective. Test that hypothesis. Did any salesperson blame themselves? Or did they all blame a different group in the organization? Did they maybe, just maybe . . . blame the product?

Sales teams can fan the flames leading to the famed Second System Syndrome sketched by Fred Brooks in *The Mythical Man Month*. Unnecessary, expensive, complex, and frightening feature explosion. Ask them what they would do about all that.

"Many programmers believe themselves to be talented designers.
In fact, this is often true, but there is a difference between design-
ing for function and designing for humans."

   —Alan Cooper, *The Inmates Are Running the Asylum*

CATEGORY 9: ABILITY TO CHANGE

"We watch our competitors, learn from them, see the things that
they were doing for customers, and copy those things as much as
we can."

—Jeff Bezos, CEO Amazon.com, quoted in
Fast Company magazine, August 2004

Every one of the six hundred-plus technology companies with market capitalizations above $200 million has done at least one thing *really* right to have gotten to where they are today. But they have also done a number of things not so right or flat-out horrendously, and it's crucial to know *how they handle failure.*

In some environments creativity ceases. Out-of-the-box thinking ceases. Technologists dominate and always have and any hint that technologists don't know user experience and suffering is muzzled out. Such places are very scary environments. They exist all over. You don't need to look far.

> *What is the biggest change you have instituted in the organization in the last three years?*

Tech organizations—you might incorrectly assume—must be changing all the time to react, proact, or just act in an industry of substantial uncertainty and turbulence. It's scary to realize just how many have environments characterized by a lack of change.

> *What was your biggest failure in past product creation?*

It's clearly a *terrible* answer if management can't think of anything they did that failed.

If management has a tendency to fire people in the wake of failure, you may want to make sure you haven't stumbled upon a fail/fire culture, since that is a good way of stopping *anyone* from suggesting something out-of-the-box.

While you're at it, ask them the last two questions from above. *What great idea on customer experience have you copied from another company?* And *How quickly did you recognize a product mistake and how did you overcome it?*

> *"Only those who risk going too far can possibly find out how far one can go."*
>
> —T. S. Eliot

> *"The dominant culture in most big companies demands punishment for a mistake, no matter how useful, small, invisible."*
>
> —Tom Peters, *In Search of Excellence*

Did anyone get fired at Apple for the Cube or the Newton?

CATEGORY 10: LEARN-RELEARN

> *"The top performers create a broad uplifting, shared culture, a coherent framework within which charged-up people search for appropriate adaptations. Their ability to extract extraordinary contributions from very large numbers of people turns on the ability to create a sense of highly valued purpose."*
>
> —Tom Peters, *In Search of Excellence*

More touchy feely stuff going on here and a great way to round out the most intensive and perhaps bizarre set of questions management thinks it has ever experienced.

There are two types of organizations. There are those that operate such that when someone fails they punish them or, better yet, fire them.

This leads to a very frightened and risk-averse organization. Most people see that this is a bad idea but many corporations are up to their eyeballs in this mentality and a miracle would be required to alter the demeanor. Then there are "learn-relearn" companies. In those companies it is understood that individuals will fail as part of the process of the entity growing to greater and greater capability. Employees learn, fail, relearn, fail, relearn, and grow ever smarter along the way. Technology sage Esther Dyson once said, "Always make new mistakes." In "fail-punish" environments, employees tend not to be around to make a new mistake. That's not *good*.

If in the process of asking management how they disrupt their own organization, their heartbeats didn't quicken now and again, well, *terrible* answer. They probably don't get the Change Function.

If there is a cartoon bubble over their heads that reads *All I ever wanted to be was a simple technologist, and here I am instead reliving the Spanish Inquisition,* you may have the wrong person for the job.

> *Is it more important to focus on increasing a user's perceived crisis or lowering their TPPA?*

No right answer here . . . could go either way.

It isn't about making technology available. It is about making technology relevant, and it is about making that relevant technology as painless to adopt as possible.

"The important thing is not to stop questioning."

—Albert Einstein

TWELVE

WHAT TO DO?

> *"The significant problems we face cannot be solved at the same level of thinking we were at when we created them. . . ."*
>
> —Albert Einstein

The primary goal of the Change Function model is to open up the possible prescriptions for addressing technology's current malaise. I want to show that we can break free of the constraints of the older, supplier-centric model and the engineering dominance that accompanies it.

To review, then:

**Supplier-centric adoption model =
f (Andy Grove Law of 10x disruptive technology × Gordon
Moore's Law)**

In that context, the list of To-Dos is quite straightforward:

1: Grove's Law → come up with 10x technological changes
2: Moore's Law → get cost down to that mystical magical tipping point

And again, in that context, the question of *who*, exactly, has stuff To Do is equally straightforward:

1 → technologists and engineers

2 → technologists and engineers

And so, in the supplier-centric model, the magicians—technologists and engineers—participate in the search for change, but no one else matters that much.

When we introduce the Change Function—

$$Change\ Function =$$
$$f\,(\text{user crisis vs. total perceived pain of adoption})$$

—a perfectly reasonable fear might be that it is equally constraining, that a shift to a user-centric model might focus on the user to the exclusion of all else, including those once powerful technologists and engineers.

That's not the purpose of the Change Function at all.

The implications of the old model were limited but not wrong! They were not sufficient but they are still necessary! Without the magicians, nothing happens. The goal is to connect the magicians to the experience of the end users.

The To Dos of the Change Function are to ask the questions at its core:

1: What can be done to *increase* User Crisis?

2: What can be done to *reduce* Total Perceived Pain of Adoption?

That's it. The ultimate goal of the Change Function?

Figure out what people *really* want!

I know what you're thinking: "Duh! Of course we want to figure out what people *really* want!"

Through the course of this book, I hope I've been successful in conveying that, as obvious as it may seem, it is precisely that—figuring out what people want—that technology companies have consistently failed to do. Those who do it right: the folks running places like Starbucks and Anthropologie and Amazon and Apple. They've long been riveted on creating user experiences that satisfy us in as many ways as possible. And as startling as it may seem, technology companies have not! Technology companies have locked themselves into a mind-set of what they *can* supply—not what they *ought to* supply.

How did we get so lost?

Because those in the technology ecosystem think they are covering the traditional bases of due diligence—employing surveys or focus groups or calling on anthropologists and consultants—but in reality they aren't. More often than not, engineers and developers have already decided what is going to market before they ask the questions. That, or the surveys, focus groups, and the few anthropologists that are actually utilized are insanely bad at figuring out what people really want.

You might say that the only missing ingredient is to let all the consultant and anthropologist types do their jobs in figuring out what users really want—as was the case at Reactrix—as opposed to being captive to what someone wants to hear and that is partially on target. It may take an additional generation to bring them out of their shell but companies that can reduce their need to hear "correct" conclusions are at an advantage straight off.

A lot of people do sit around and review poorly constructed surveys and the findings from mediocre and misdirected focus groups. The industry's indentured anthropologists are merely yes people who have been selected in the very same way that expert witnesses tend to be selected in courts of law. And in this case, it's the anthropologists and consultants that subconsciously influence technologists and managers just as so-called experts can influence a jury.

In the most extreme situations, they will even *buy*—literally—consulting reports that will tell the world about the magic just created and how it will be a big, big market:

"You always see gushing analyst white papers as part of the press package when a new product comes out. We needed one, too. Well, it turns out that you buy them. We negotiated with several firms. Aberdeen wanted $16,000, preferably in stock, but Seybold would do it for $15,000 in cash. So we used Seybold, who sent us a nice man who wrote exactly what we told him to. I didn't like it, but everyone made it very clear that this was standard practice, that it was necessary for us, and that I should keep my mouth shut and stay out of the way. So I did.

—Charles Ferguson, *High St@kes, No Prisoners*

Whatever way it happens, a "common wisdom" inevitably develops that may be devoid of common sense. People quit digging for insight and truth and submit to what has somehow become established thought. And fighting the common wisdom of or the group think within a company is usually a course too painful to pursue.

Steve Kemper's *Code Named Ginger* followed the development of the ridiculously ingenious *Segway*—the self-balancing people mover of sorts. You may have seen one of these machines or maybe even three or four of them. But you haven't seen one hundred million of them, as some early predictions led us to imagine. Sure, they were *cool*. But feedback from so many other angles suggested the Segway would never quite catch on in a truly meaningful sense. All that stuff was ignored.

Ignored

The idea, for example, that the Chinese people would abandon automobiles because the Segway solved societal issues of pollution, energy shortage, and intense urban traffic patterns was far more persuasive than whether or not this obviously smart technology would really appeal to the individual. The power of the engineers', developers', and investors' group think about the technical genius of the product overwhelmed all else. The technology *had* to fit somewhere, so they piled into this ill-conceived product and drove a production schedule completely divorced from realistic expectations.

What's a better way to approach what users *really* want that goes beyond surveys that tell us what we wanted to hear—including those in which the terms of the contract actually *assured* that we would hear it? I propose an exercise that might be called Find the Missing Insight.

Let's suppose there are three primary types of relevant inputs in our search for answers:

- Data
- Information
- Insight

They Are Not the Same!

Data is numbers and the like.
Information is data in context.
Insight is discerning the true nature of a situation.

The first two aren't that hard to find. What we *really* want is insight as to why someone will buy and use or *not* buy and *not* use a potential product.

When you conduct a survey or run a focus group, you collect *data*.

You can certainly gather a lot of data. But so what? Can data alone help you successfully determine whether a product will succeed? Not at all—especially if you haven't turned that data into valuable information by finding meaningful context.

The problem is, the *context* of surveys or focus groups is a laboratory of sorts as opposed to real world experiences. What we *say* in a focus group may or may not resemble what we would *do* in the real world. And that tends to be the result of faulty premises. Pepsi continually won the Pepsi Challenge but failed to gain market share. Why? Because although the challenge found that people favored sweeter drinks in small amounts; in the real world, people prefer less sugar if they are drinking a whole can or buying a whole case. There was nothing insidious about the challenge, but for the most part, we are all terrible at designing prefabricated replicas of the real world.

And then there is the people part.

As humans, we're all somewhat half baked. Part of the human experience may be that we pretend to live more rationally than we do. Or, its opposite—pretend to live more emotionally than we do. Or more like who we wish we were than who we actually are.

If the only information we come up with is what humans would do in the fake world of surveys and focus groups, are we any where near insight? Not really. Insight, again, is true clarity in explaining a phenomenon.

We want insight not data.

Recall Ted Levitt of Harvard's famous observation: People don't want quarter-inch drill bits—they want quarter-inch holes.

Folks buy a service. An insight. Folks buying their kids milkshakes at McDonald's are buying peace and quiet in their cars as their children struggle but persist in consuming the thick sugar-laden substance. Can that insight be found in the data? No. Can it be found in the information? No. Insight is beyond data and information.

Insight is the glue between technology and users.

The Change Function is a framework in which to uncover insight in a systematic way by continually asking about users' crises and their total perceived pain of adoption. The model takes information as input and delivers insight as output by asking two simple questions:

> 1: What can be done to *increase* User Crisis?
> 2: What can be done to *reduce* Total Perceived Pain of
>   Adoption?

There are many that prefer to think that the commercial success of technologies is unpredictable. They like to say, *Well, you never* really *know how technology will be used. . . . Remember when we invented x, and we thought it would be used for y but, lo and behold, when we got it into the market, folks used it for z!* This thinking runs pretty deep. It's called a solution looking for a problem.

> *"I prefer 'technology push'—find an interesting new technology*
> *and try to come up with uses for it. 'A solution looking for a*

problem' is supposed to be a bad terrible epithet, but in my expe-
rience it works."

—Legendary integrated-circuit designer Carver Mead, quoted in
MIT's *Technology Review*, September 1, 2004

Why is the idea of a solution looking for a problem so widely accepted? Because urban legends of unintended uses support a notion that
we should just create as much fresh technology as possible since it is impossible to know the ultimate end user applications! That's a cool idea
for technologists and engineers. It also provides a great justification for
failure: *Hey, we created cool technology, but the key application never
showed up, which of course isn't our fault.*

The idea that there's a "lottery" aspect governing development, and
the reality of the marketplace suggests there isn't a better system. It's certainly not a very scientific mode of operation—which in technology is
particularly odd, since it is scientists themselves who seem willing to
adopt this thinking about the unpredictability of applications for technology. The implication is there is nothing "to do."

**The technology industry should not—and cannot—consist of
solutions looking for problems.**

Why? Because it's a very bad idea. For every instance that a solution
actually finds a problem there are a-million-and-one orphaned products that we fail to remember—when our selective amnesia goes to
work.

Nothing to do? Not in the slightest.

TO DO 1: EMPLOY RAPID PROTOTYPING, CONTINUOUS ITERATION, AND CODESIGN.

*"Healthy companies know that they have to allow people to fail
without assessing blame. The overarmored organization has lost
the ability to move and move quickly. When this happens, stan-*

dard process is the cause of lost mobility. It is, however, not the root cause. The root cause is fear."

—Tom DeMarco, *Slack*

What if we failed early—and failed often—how might that help?

In other words, what if we came to understand a user's "Crisis" and "Total Perceived Pain of Adoption" by creating a process where we failed and iterated and failed and iterated, again and again? If we created a system of trial and error that provided *many* opportunities to see what did and did not sync with the user.

In the simplest sense, rapid prototyping entails shrinking the time between idea creation and the crafting of a prototype that can be assessed and perhaps actually used by real world people in real world situations. The key: an acknowledgement that the product is expected to fail.

Why does group think form? Why do "yes people" lean toward saying what they think the boss wants to hear? Because the environment for saying the opposite is too dangerous. No ones wants to say a product stinks if the boss has decided to go ahead and already has four million of them sitting on the loading dock. At that point, it's too late.

What if it was never too late to say a product stinks and if saying so was actually encouraged—as opposed to disingenuously encouraged? Rapid prototyping and continuous iteration establish exactly that environment.

Most organizations are run in the following way:

Fail → **Punish!**
Or a variant known as
Fail → **Fire!**

In those environments, it becomes a big deal to say a product flat out "stinks"—because heads will roll. As a result, bad products get through the system unnaturally and create unsuccessful results. But what if prototypes were expected to fail early and often and continuous iteration was the norm?

The culture would shift to:

Learn → Fail → Relearn
or
Fail Fast and Iterate!

How does all this tie back to data, information, and insight?

Well, the requirement to come up with one **major** insight drops considerably. There is no longer a need for an epiphany. Successive rounds of trial and error can provide the data—and, more importantly, the information—that leads to a series of ever-so-modest adjustments.

The expectation that a product will fail frees an organization to actually think about what needs to be changed—not just once, but over and over and over again—and, if your customers are empathetic to your efforts, they might even provide spectacularly beneficial codesign. They may come to own the product.

EBay has done codesign with its users so well that its "community" gets pretty ticked off if eBay proper makes changes without asking the entire ecosystem.

Amazon used the anonymity of the Internet to its advantage in making customer complaints available to senior executives on a daily basis. People often tell you what they really think when they are out of reach of physical or verbal assault.

Salesforce.com releases five to ten more iterations of its software than its larger, more established peers.

The key aspect of the Learn → Fail → Relearn approach is that development is never truly complete. Never! It's fine to celebrate improvement now and again, but keep it short, since the job is never finished!

This brings us back to the fact that Larry Page thinks his Google search engine is really mediocre. He thinks that despite the fact that, by most measures, Google is the most effective search engine on the planet and has certainly found commercial adoration. Larry Page thinks of Google's opportunity in search as not just those times when you hit enter on a well-conceived search and seconds later get relevant results but also those failed searches and the time spent considering how to contort the human mind into the mind of his machine. Larry Page likely wishes that at the moment a search gathers in our mind—even in its crudest form—Google could help pop the result straight into our head! Now that would be a great search tool and a great business. Google doesn't do this . . . yet.

Douglas Engelbart—the creator of the mouse—is *very* depressed when he contemplates how little progress computers have made in assisting the inhabitants of Earth. While he's perfectly able to admit that computers are fantastic in many regards, he thinks they absolutely stink in a much larger sense of what creators like he and others envisioned— as a seamless frictionless symbiotic relationship between man and machine as opposed to the tense, hostile, frustrating condemnation many feel toward technology. In September 2005, late one night while working on the final pieces of this book, my waiter at the Thornwood Dinner in Thornwood, New York, complained to me that his Dell had been breaking down a lot, sheepishly adding that he was, "pretty much just a novice." I heard myself responding for me, Doug Englebart, and many other earthlings that "we are all novices . . . and it shouldn't matter!"

TO DO 2: HIRE REAL ANTHROPOLOGISTS, CREATE GUIDING "CRISES," AND FILL THE GAP.

The next step in determining what people *really* want is to employ those who make their living by studying people at a deeper level; to employ those able to see more than a series of specific responses to specific stimuli when confronted with data or information—to see the patterns of behavior in a holographic as well as in a holistic sense. Employ experts who can see how we act in ways consistent with our deepest desires and fears.

Holographic and Holistic

Who are these experts? Sociologists, anthropologists, communications consultants, change consultants, professional observers, futurists, and folks who just study change a whole heck of a lot. The famous sage Yogi Berra once said that you can observe a lot by just watching. That seems right to me.

Some of these folks apply their skills in forecasting the likely success of product launches. Many work on "user experience" as opposed to just products themselves. There aren't a lot of people with a successful confluence of experiences that are required to do this work successfully but they exist. And it's a good idea to hire and deploy them to create a value structure for your organization: a set of core beliefs about what *crisis* folks have that a particular technology might address—crises such as the need for mobility, the addiction to communication, a desire for integration and simplicity, and virtualization.

But that isn't enough. If the change-anthropologist-sociologist-futurist gurus enter stage left, drop in some guiding crises, and then exit stage right, the value will be lost. Technologists and engineers will return to creating cool technologies and a single-minded focus on cost reduction as the be all and end all of product development. In other words, a total waste of money.

Change experts have to be planted inside the organization. And they had better be protected, otherwise they'll be hijacked, or worse. Before long, they'll find themselves on stage at a Wall Street analyst meeting saying nice things—incomprehensible as those things might be to financial types—so that the boss looks good in connection with the new products that are about to be released with great fanfare. Or they'll be sidelined by those seeking to protect their own turf in the development process. If you go to the trouble to bring them into your group to deliver insight in a world filled with mere data, you may as well go to the trouble to protect them as independent thinkers.

The good news: there's a perfect place to put them. There's currently a massive gap in nearly every organization on the planet between development and marketing. Just look at any income statement—research and development are lumped together as one line item even though there are mountains between the two. Sales and marketing are also usually coupled together, although there should be a mountain between those two as well! But the largest of the mountains separates R & D from sales and marketing.

Development types are charged with taking a piece of research and developing it into a product of some sort. Marketing folks are charged with turning what they are given into something sellable, long after the development is done. There is a gap—and those who understand the holographic and holistic tendencies of Earthlings can fill it.

Maybe we should stuff the anthropologist in the gap and maybe, just maybe, they ought to report straight to the CEO. And then we should help get their guiding "crises" embedded into development. And assist them in changing marketing from a mechanism designed to "get" or to "trick" potential users into buying whatever has been created into a servant of those potential users.

Great marketing enables personal epiphanies:

"I want this product now."

Change gurus can play a role all the way from identification of the organization's guiding "user crises" through to establishing in what ways the product actually does deliver at a deeper level—ways it delivers what Ted Levitt termed a service.

Don't expect the people in development or marketing to embrace these gurus with open arms—because adding them into the ecosystem suggests that someone hasn't been doing their job. But in reality no one *has* been doing this job of experience design to any great effect. Starbucks only recently woke us up to it with everything they do that goes beyond mere coffee.

It also a smart idea to charge an independent group of change gurus with a devil's advocacy assignment to find holes in what appear to be otherwise spectacular products. IBM used a group of such people to battle test software. They were not popular, but they were effective. Imagine a SWAT team devoted to tearing apart the software created by the development team to spot faulty code. Imagine a team that congratulated themselves when they found more and more holes, or when they found security weaknesses—when they found problems before the code even made it to a beta group. I wonder where Microsoft would be today on security if they had employed a similar SWAT team. Devil's advocates can anticipate the potholes a product line will have in its life cycle by anticipating how the demands of users might shift across a three- to five-year period. But again, they will need protection!

TO DO 3: UTILIZE TRUE OBSERVERS, TRY ON PERSONAS, AND EMPHASIZE CULTURAL CHANGE.

> *"The high-tech industry has inadvertently put programmers and engineers in charge, so their hard-to-use engineering culture dominates."*
>
> —Alan Cooper, *The Inmates Are Running the Asylum*

OBSERVERS

There's a Swedish movie from 2003 called *Psalmer Fran Koket—Kitchen Stories*—that's set in the 1950s. In it, a group of Swedes cross the border into Norway to sit in tall chairs in the kitchens of single men to observe patterns of behavior in order to develop a smarter kitchen.

It's an interesting movie, but as with surveys and focus groups, there are more useful and less useful methods of observing how people behave. "Kitchen Stories" falls in the camp of "less useful"—the single men in the middle of the wilderness come to regret the deal they originally made to be observed and, well, a lot of other stuff happens in this highly celebrated film.

Observers collect data. But data is only useful as a means to the end of generating insight. Great observers can get at the underlying crisis of what people want. Great observers can identify the pain of adoption.

Consider the example of IDEO, widely regarded as the foremost design company in technology. Stanford's David Kelley founded the firm in 1978, and its people are experts at observing folks and taking what they learn all the way to product design. In my opinion, they are masters of the Change Function—observing people very closely in order to identify user crisis and total perceived pain of adoption. But beyond IDEO, there are very few trained observers with a specific domain expertise in technology and enough formal academic and real world training. If you find one of these people, you might want to hire them.

As much as we hear words like *user experience* and *simplicity* and *ease of use* in today's environment, there are very few people in the technology ecosystem that have any idea what those words actually mean. Saying simplicity does not mean being able to create simplicity.

The people at IDEO are superb at product design because they seemingly know each and every little thing that might irk a user, and they design accordingly. Do you ever wonder why Apple has been able to create an entire collection of engaging, crisp, simple, usable products with the slightest of features that enhances user experience by the old "order of magnitude"? Like the flywheel on the iPod? The photo iPod? Or the impossibly small iPod nano? It isn't an accident. Apple has developed a culture that's sensitive to the ways people actually like to use technology products!

Do you ever wonder why the same companies continue to make generation after generation of dumb smartphones that exude complexity and fail to hit sales expectations? Or why we all have six highly complex remotes floating around our living rooms? How many remotes does it take to change a channel? The people that make them aren't studying people. They respond to commercial failure with even greater complexity, thinking that there must be a tipping point at which a fully loaded version will gain broad acceptance. Do you ever wonder what planet these people live on when they're not working at their day jobs in Silicon Valley?

But there *are* experts who really understand nearly every element of the user interface. Shortly before his untimely passing, Jef Raskin, creator of the Macintosh, wrote *The Humane Interface* to get across everything he has contemplated about simplifying user interfaces. Jef believed—like many other user interface experts—that there should be a morality with regard to interfaces that would not limit a potential technology to those who have had more education with technology across a lifetime and are able to utilize an additional technology quite easily while those without such a background are unable to do so. According to this grand vision, every new technology should be universally and completely accessible to every individual no matter what background.

A moral argument.

PERSONAS

There's a new buzz in the user experience world—it's called *customer centricity*. A number of companies are establishing programs that employ it in one form or another.

> *"We are in the midst of a strategic transformation to put the customer at the center of all that we do," said Brad Anderson, vice chairman and chief executive officer, speaking ahead of the company's annual analyst meeting. "Our customer centricity initiative enables us to engage more deeply with customers by empowering our employees to deliver tailored products, solutions and services to customers through our stores, Web sites, call centers and in-home services. This work has allowed us to look at our business through a new lens. As a result, we see additional opportunities to connect more closely with our customers, to increase our competitive advantages, to boost our market share and to deliver superior financial results."*
>
> —Best Buy press release, May 2004

What is Best Buy *really* doing? What's all that nice chatter mean? Best Buy is altering its culture from being supplier-centric toward being customer-centric by the mere act of bombarding its own employees with the same message each and every day. Best Buy is re-creating its culture by planting "customer centricity" in its daily life. Big stuff!

Best Buy is creating personas. In order to understand their customers far better than they ever did before, they've crafted five personas of customers they believe routinely visit their stores. In doing so, the company is aiming to be far more effective in creating separate and successful holistic experiences for those different groups.

The "personas" as described by Best Buy in their own press releases:

- Barry—The affluent professional who wants the best technology and entertainment experience and who demands excellent service.
- Buzz—The focused, active, younger male customer who wants the latest technology and entertainment.
- Ray—The family man who wants technology that improves his life—the practical adopter of technology and entertainment.
- Jill—The busy suburban mom who wants to enrich her children's lives with technology and entertainment.
- BB4B—The small business customer looking for solutions and services to enhance the profitability of his or her business.

Best Buy created personal assistants to help the suburban moms and developed the Geek Squad—a twenty-four-hour support mechanism—for small business owners.

What have the results been? Comparable store sales were 7 percent higher at thirty-two initial "lab" stores than at the balance of U.S. Best Buy stores. The percent of shoppers actually buying products was 6 percent higher at the lab stores than at the others. Best Buy expanded the program quickly.

Alan Cooper, author of the 1999 book *The Inmates Are Running the Asylum*, is in a select group of folks that have superbly described the lack of relationship between the creation of technology, the development process, and the actual user experience. Back in 1999, when he described the use of personas, it was somewhat controversial, due to its seeming disrespect for the current methodology of the time.

The use of personas is just one of many strategies for bridging the gap between technology creation and user experience. What Cooper proposes is similar to the Best Buy enactment but a little bit richer. Key participants in the technology creation process—both designers and programmers—take on a persona for a month or more at a time while working with his team of consultants. Members of the team refer to each other by new names representing their personas; it's a full immersion process.

An example: "Chuck Burgerweister, business traveler. A 100,000-mile club member who flew somewhere practically every week. Chuck's vast experience with flying meant he had little tolerance for complex, time-consuming interfaces, or interfaces that condescended to novices."

"Chuck" was developed when Alan Cooper's group was working with Sony to develop a back-of-the-seat user interface for on-board entertainment systems. The team then asks questions: "What is Chuck *really* like?" and "What matters to him most of all?"

Members of the development team are expected to insightfully play the role of their persona. In such exercises, the development team increases its empathy for the plight of the user to a far greater degree than that derived from mere off-sites or brain storming sessions. Members of the team are steadily transformed into advocates for their personas. Someone actually becomes "Chuck."

Personas help designers and developers see the possible crisis and the total perceived pain of adoption in ways not easily gathered by untrained observers. They capture *insight* as opposed to collecting *data*.

If you are able to lead your design and development team into truly inhabiting their personas, great results can occur. Insights pop out when staffers track their persona's emotions in relationship to the product at hand. Surveys and focus groups represent a world that doesn't exist. The persona's world doesn't really exist either, but is more effectively conjured during every moment that individuals are playing their roles.

WHAT DOES IT ALL ADD UP TO?

Taken together, the mandates engendered by the Change Function offer the beginnings of an altered organizational structure and philosophy:

- Fast Fail: *learn → fail → relearn → iterate* replaces *fail → fire.*
- Codesign with users → iterate to reduce the magic normally involved in figuring what they want.
- Employ actual experts in assessing user crisis and total perceived pain of adoption.
- Create guiding crises for your organization to reference like a compass.
- Protect experts inside the structure and insert them early in product establishment.
- Employ a protected devil's advocacy process.
- Hire professional observers.
- Integrate professional observation early in the design and development process.
- Develop customer-centricity programs to shift culture.
- Use strategies, such as personas, to materially change the effectiveness of design and development.

If you really want to upset the apple cart along the way—to go for the whole enchilada—you may wish to attempt the following cultural shift. But be forewarned: it should come from the top and the degree of difficulty is immense.

Change the language.

Change the language from a supplier-centric one to a user-centric one.

Rid your organization of supplier-centric terms like *sales cycle* and replace them with user-centric alternatives such as *buying cycle*. This will throw people into a tizzy because it indicates there's a new sheriff in town. But remember what Machiavelli said: the new order will find few supporters at first—so don't do things half-heartedly.

While you're at it, you might also condemn all forms of water cooler chatter complaining about customers. In doing so, you'll run the risk that "they" may pour the water cooler over your head—and not in that you-just-won-the-Super-Bowl! kind of way. But the risk is worth it, depending on the culture you dream of and the results you wish to create. There is an undercurrent in the technology world of blaming customers, of considering them stupid for not "getting it." That culture needs to be softened, if not broken, for a user focus to push out the supplier orientation.

> *"All truths go through three stages. First, it is ridiculed; second it is violently opposed; and third, it is accepted as self-evident."*
>
> —Schopenhauer

> *"First they ignore you, then they laugh at you, then they fight you, then you win."*
>
> —Mahatma Gandhi

The purpose of this book is really quite simple. I think there's a problem, and I'm proposing a solution. The problem? The technology industry has become self-absorbed as a result of five decades of success. (Who wouldn't?) Over time, technologists have become increasingly focused on creating miracles, even if it's rare that those miracles translate into commercial success.

In late 2005, I attended a black-tie tribute to Gordon Moore at the Waldorf Astoria. What a smart, kind, caring, engaged, and interested man. There was a sense that evening that Moore and many of his peers in the 1960s were creating their magic to make the world a better place. Building successful, enduring businesses was a critical component in that, but it's obvious that Moore cared deeply about users as he laid the groundwork of technology for future generations.

Moore's Law has endured to affect millions of technologists around the globe since Moore's legendary article in a 1965 issue of *Electronics*, but it's not entirely clear that his spirit of making the world a better place through technology has itself endured. On that night of tributes, recent technologies were lauded for having changed or transformed the world. But nobody had the courage to suggest that those technologies had made the world a better place.

And it isn't clear that the world has become a better place. It isn't clear at all if technology is pulling its weight. Transformed? Yes. Changed? Of course. Better? Not sure at all.

The world sure seems to need all the help it can get. It needs people to listen to each other and it could use technology's miracles along the way. What *The Change Function* says at its heart is that you will only be accidentally successful if your focus is on what *you* can create. Systematic success—whether it's in creating new products, building great companies, or changing the world—comes to those who manage to see the world through the eyes of others . . . to understand their crises and to help them find less painful ways of changing their world for the better.

INDEX